M. Osteaux
K. De Meirleir
M. Shahabpour
(Eds.)

Magnetic Resonance Imaging and Spectroscopy in Sports Medicine

With Contributions by
E. Achten · T. Buisseret · P.-P. Casteleyn
R. Luypaert · K. De Meirleir · M. Osteaux
C. Pierre-Jerome · M. Shahabpour · T. Stadnik
Y. Taeymans · P. Vaes · M. Van Cauteren
K. Vandenborne · P. Van Roy · E. Verhaeve

Foreword by H. G. Knuttgen

With 144 Figures
in 244 Separate Illustrations

Springer-Verlag
Berlin Heidelberg New York
London Paris Tokyo
Hong Kong Barcelona

ISBN 3-540-52548-3 Springer-Verlag
Berlin Heidelberg New York
ISBN 0-387-52548-3 Springer-Verlag
New York Berlin Heidelberg

Library of Congress Cataloging-in-Publication Data. Magnetic resonance imaging and spectroscopy in sports medicine / Michel Osteaux, Kenny De Meirleir (eds.); with contributions by E. Achten ... [et al.]. p. cm. Includes bibliographical references. ISBN 3-540-52548-3 (alk. paper). – ISBN 0-387-52548-3 (alk. paper). 1. Magnetic resonance imaging. 2. Nuclear magnetic resonance spectroscopy. 3. Sports medicine. 4. Sports – Accidents and injuries – diagnosis. I. Osteaux, Michel, 1944-. II. De Meirleir, Kenny. III. Achten, E. (Eric) [DNLM: 1. Athletic Injuries – diagnosis. 2. Magnetic Resonance Imaging. 3. Nuclear Magnetic Resonance – diagnostic use. QT 260 M196] RC1225.M34 1990 616.07'548 – dc20 DNLM/DLC for Library of Congress 90-10399 CIP

Typesetting: Fotosatz-Service Köhler, D-8700 Würzburg
21/3130-543210 – Printed on acid-free paper

Foreword

Throughout the twentieth century, interest in sport has evidenced a major increase worldwide and a new field of sports medicine has emerged. Sport spectatorship has been stimulated in part by the increased ease of international travel and the introduction of new means of communication. Interest in professional sports and in such events as the World Cup and the Olympic Games has never been higher and the general popularity of sports has drawn increasingly greater numbers of young people to sports participation.

Interest in sports participation has been strengthened both by the provision of numerous sports facilities to populations throughout the world and the accumulating information base that relates regular participation in programs of sport and exercise to health enhancement. Along with increases in interest and participation has come a need for a better understanding of the physiological changes accompanying muscular activity and of the trauma that can result from both competitive sport and recreational exercise.

The term "Sports Medicine" has come to include sports traumatology, exercise physiology, and sport biomechanics as well as involve many aspects of such areas as environmental physiology, clinical cardiology, primary care medicine, orthopaedics, and physical therapy. Scientists and clinicians of these various areas have been drawn together for cooperative and collaborative projects in the assessment of persons engaged in exercise and sport. Both groups are highly dependent on the measurement techniques and equipment available to assess the morphologic and metabolic status of cells, tissues, and organs as well as the responses and adaptations of the human organism to environmental stimuli.

The development of new technologies and associated instrumentation during the last half century has facilitated both the acquisition of basic information regarding responses and adaptations of tissues and organs to participation in exercise and sport and to the identification and understanding of associated injuries. The introduction of the technique of magnetic resonance imaging has provided a powerful new tool for the scientist and the clinician.

The opportunity to examine tissue morphology and metabolite concentration with no surgical involvement makes the technique highly valuable to the researcher and clinician alike. The scientist can better address questions related to conditioning for competition, athlete performance, fatigue, and recovery while the efforts of both the clinician and the scientist are facilitated regarding the advancement of knowledge regarding injury, disease, healing, rehabilitation, and the role of conditioning in injury prevention.

In this text, Prof. OSTEAUX, Prof. DE MEIRLEIR, Dr. SHAHABPOUR, and their collaborators have provided us with a wealth of information regarding the use of magnetic resonance imaging with special relation to applications in sports medicine. In addition to providing the reader with complete and authoritative coverage of the topic, this information will constitute a powerful stimulus to the identification of new applications for magnetic resonance imaging.

Prof. HOWARD G. KNUTTGEN, Director
Center for Sports Medicine
The Pennsylvania State University
University Park, Pennsylvania, USA

Preface

The past two decades have seen a major evolution in the world of sports, together with spectacular innovations in the diagnostic arsenal of sports medicine. Without doubt, magnetic resonance represents a major diagnostic advance in medicine. Its role in sports medicine, however, is still evolving, and a number of indications and applications are only beginning to become apparent.

The safety of this technique in humans is well documented and provides probably the strongest explanation for its rapid development and its importance in sports medicine. NMR does not involve ionizing radiation, and studies have revealed no deleterious effects of magnetic fields.

Today the work-up of patients with sports-related musculoskeletal trauma often includes MRI. It appears that early diagnosis of avascular necrosis will be facilitated by MRI. In athletic injuries of the knee, MRI is strongly indicated for differentiating partial and complete tears of the cruciate ligaments of the knee. Here, MRI will replace arthrography in the future as soon as this technique becomes available in more clinics. Other indications for MRI in sports-related injuries include shoulder and ankle ligamentous injuries, collateral ligament injury at the knee, meniscal injuries, and possibly in the future even soft-tissue trauma and hematoma.

MRI greatly augments our diagnostic arsenal in the detection of central nervous system lesions, which in some instances is of utmost importance for sports medicine. The internist is often confronted with the problem of exertion headaches. Although these are often benign in nature, they must be examined carefully for underlying disease. Related to this problem, MRI allows noninvasive diagnosis of subdural hematoma, arteriovenous malformations, and occasionally aneurysms. It has the capacity to determine the presence of potentially lethal lesions, prior to hemorrhage.

With the advent of ^{31}P nuclear magnetic resonance it is now possible noninvasively and continuously to monitor pH and intramuscular metabolism and to study exercise-induced muscle injury. One of the principal obstacles that has limited studies of energy metabolism in exercising human muscle has been the invasiveness of the techniques used, such as tissue biopsy and subsequent biochemical metabolite assay. For exercise physiologists NMR has opened a new world for the study of muscle performance; for the first time, detailed examination of the relation ships among muscle fatigue, pH, and phosphates is possible. This is discussed in detail here.

As it is possible to obtain ^{1}H NMR spectra of muscle tissue, it will eventually become possible to apply this technique to the study of lactate production in exercising muscle. Also, the noninvasive measurement of intracellular glycogen and glucose will be possible with ^{13}C NMR, which also will open new horizons for the exercise physiologist studying glycolytic metabolism.

Cardiac MRI offers an excellent opportunity to investigate cardiac morphology in trained athletes. Furthermore, it is useful in sports medicine to obtain information concerning cardiac function (ventricular dimensions, ejection fraction, and left ventricular wall thickening) and to image the various changes in cardiac structures which may be sports related.

This book begins with a short and very simple physical and technological introduction to magnetic resonance imaging and spectroscopy. The main objective of the work – and the largest part of the book – is devoted to the familiarization of the relevant specialists with the proven clinical indications, contributions, and semiology of magnetic resonance in sports medicine. This objective is pursued in applications of MRI to the osteoarticular system where this is technically feasible and clinically interesting (knee, ankle, hip, shoulder, elbow, wrist, and spine). These various topics are treated with a depth and detail corresponding to the importance of the clinical contribution of MR to sports medicine. In particular the knee joint – the most commonly injured joint in athletes and the area in which the advantages of MRI are most spectacularly apparent – is discussed in detail and is abundantly illustrated.

The third part of the book deals with more novel ideas, still in a phase of clinical research, and MR tools for the study of sport physiology. Application of MRI to cardiology and dynamic (^{31}P) MR dynamic spectroscopy of muscle fall into this category.

Other uses of MR in sports medicine can be anticipated (or have been preliminarily reported), namely the study of body composition, perfusion-diffusion, and circulation factors in sport physiology, including, for instance, the proton spectroscopy of muscles. These topics, however, have been judged too preliminary and too distant from clinical application at the present time to be incorporated here. Clinical sports medicine, as well as sport physiology, has already been greatly influenced by the development of NMR, and advances will continue in the coming years.

We would like to express our gratitude to team of technicians of the Magnetic Resonance Unit, under the direction of Mr. W. Deplecker, for producing all imaging and spectroscopic examinations. Our thanks also to Mr. E. Broodtaerts for producing the figures and to Mrs. J. Lemahieu for preparing the manuscript.

This book is the fruit of an intensive and friendly collaboration between the team of those in sports medicine – who actively participate in most of the examinations – and the dedicated MR team of our University. We thank every member of the two teams for their enthusiastic collaboration.

M. Osteaux
K. De Meirleir
M. Shahabpour

Contents

Contributors

All of the contributors are members of the staff at the Academic Hospital, Vrije Universiteit Brussels, Laarbeeklaan 101, B-1090 Brussels, Belgium

Prof. Dr. Michel Osteaux
Professor of Radiology, Chairman, Radiology Department and MR Center, Director Pluridisciplinary Institute for Research on Imaging Sciences

Prof. Dr. Kenny De Meirleir
Professor of Sports Medicine and Human Physiology, Director of the Center for Sports Medicine, Chairman Departement of Exercise Physiology

Dr. Maryam Shahabpour
Fellow in Bone and Joint Radiology, MR Center

Dr. Eric Achten
Fellow, MR Center

Dr. Thérèse Buisseret
Fellow in Neuroradiology, MR Center

Dr. Pierre-Paul Casteleyn
Senior Staff Member, Orthopedics Department

Prof. Dr. Robert Luypaert
Associate Professor of Applied Imaging, Director MR Center

Dr. Claude Pierre-Jerome
Fellow, MR Center

Dr. Tadeusz Stadnik
Head Neuroradiology Unit, Radiology Department

Dr. Yves Taeymans
Staff Member, Cardiology Department
and MR Center

Prof. Dr. Peter Vaes
Kinesitherapy and Revalidation Department

Marc Van Cauteren
Research Fellow
in Experimental Magnetic Resonance

Krista Vandenborne
Research Fellow in Sports Medicine

Prof. Dr. Peter Van Roy
Postdoctoral Research Fellow,
Biomechanics Department

Dr. Eddy Verhaeve
Fellow, Orthopedics Department

Basics

1 Principles of Magnetic Resonance Spectroscopy and Imaging

R. LUYPAERT and M. VAN CAUTEREN

1.1 What Is Nuclear Magnetic Resonance?

1.1.1 Nuclear Magnetism

For biomedical and biochemical purposes we may consider the world as being built up of atoms and molecules. Thus, a technique permitting us to probe at the atomic scale can provide an insight into what is happening at this very fundamental level. Up to now there has only been one technique which can perform this task totally noninvasively and without any harm to the subject: nuclear magnetic resonance (NMR). This uses the rather strange property of some nuclei to behave like tiny magnets, a property called nuclear magnetism. A direct consequence of this is that their behavior is influenced by external magnetic and electromagnetic fields, just as the earth's magnetic field forces the magnetic needle of a compass to align along its field lines and thus point northwards.

NMR is based on the same principle. When an object, for instance, a patient, is placed in a strong magnetic field, the nuclei with the above-mentioned property, also commonly called spin, tend to align with the applied magnetic field. There is one difference to the case with the magnetic needle however: since we are now intervening at the atomic-microscopic level we must take the laws of quantum mechanics into account. According to these laws the nuclear magnets have clearly determined energies, and the number of spins, having a certain amount of energy, is determined by thermodynamics. As a result the spins accommodate themselves in an arrangement called the equilibrium state. Together, all the microscopic magnets cooperate and lead to a macroscopic, measurable magnetization.

As long as only a static magnetic field is applied, nothing changes this very cosy configuration. However, when the patient is subjected to a radio frequency pulse, the nuclei are kicked around and pass into a nonequilibrium state, where they do not feel as comfortable, being burdened with an extra amount of energy which they received from the radio pulse. They can relieve themselves of this stress through a mechanism appropriately called relaxation. During this process they get rid of the energy through radiation. It is this radiation (i.e., radio waves), which can be detected with very sensitive and specially adapted antennas, that gives rise to the NMR signal.

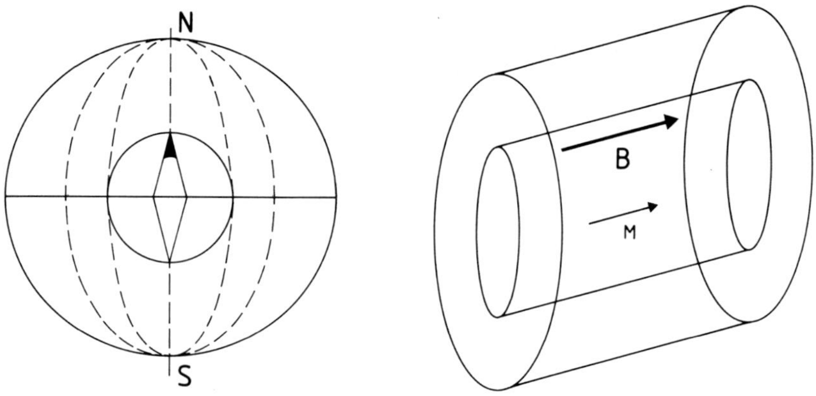

Compass in the earth magnetic field   Magnetization in the magnet

Fig. 1.1. A magnetic compass needle is forced to point to the (magnetic) north pole. In the same way, nuclear magnets, or spins, align either along the magnetic field lines or opposed to that field

Now things become really interesting. Since all the mechanisms described are the result of interactions between atoms, the signal emitted is specific to each atom. In this way, one can identify atoms, extract information concerning the chemical and physical environment, and measure the relative concentration of the respective atoms. It even allows us to make images, using some clever tricks.

1.1.2 Resonance

There is a very sharp restriction on the frequency of the pulses used in NMR experiments. The frequency is proportional to the applied magnetic field B_0 and is determined by the famous relationship first established by the physicist Larmor:

$$\omega_0 = \gamma B_0$$

where ω_0 stands for the frequency and γ is a proportionality factor called the gyromagnetic ratio. This ratio is different for each nucleus; thus we have different frequencies for the same magnetic field for protons (i.e., hydrogen nuclei), carbon, fluorine, phosphorus, etc. The phenomenon of nuclei being susceptible only to a precisely determined frequency is called resonance. It is very well known to all of us, occurring frequently in daily life. For instance, when a bell is struck, it emits a sound with a certain frequency determined by the shape and the material of the bell; another bell, with a different shape, has a different frequency. Important, however, in this respect is that a bell also tolls if another bell, this time with the same frequency, sounds; it does not react at all if

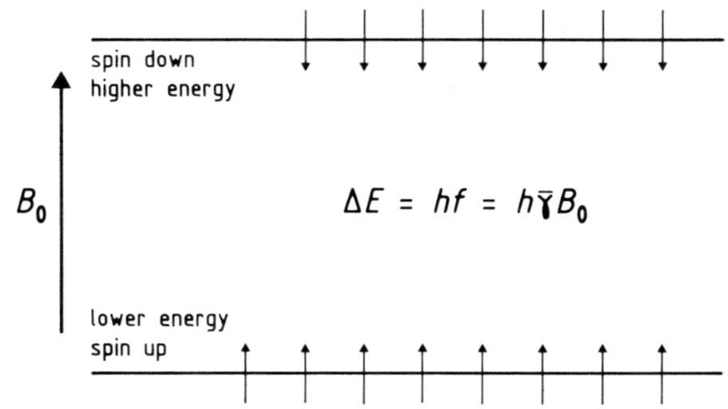

Fig. 1.2. Spins immersed in a magnetic field can occupy two positions. They can be aligned along the field lines, in the so-called up state, or they can be opposed to the field, in the down state. The latter has more energy than the former. The energy difference between the two states (ΔE) is equal to $h\bar{\gamma}B_0$ or, equivalently, hf. h is Planck's constant 1.34×10^{-34} Js) and $\bar{\gamma}$ is the gyromagnetic ratio. f is the frequency of the resonant radio waves. The frequency f is thus equal to $\bar{\gamma}B_0$, expressing that resonance is determined by the field (B_0) and the nucleus itself ($\bar{\gamma}$ is specific to its nucleus)

the other bell has another frequency. Thus, we can activate a bell from a distance if we excite it with a sound with exactly the same frequency, or, in other words, if the two bells have the same resonance frequency. The same is the case with our nuclear magnets: they react only to their own resonance frequency.

To understand resonance we must use quantum mechanics. In the simplest case, which is of interest for our purposes, the nuclear spins in the equilibrium state can have only two values for their energy (their energy is quantized). The lowest level has more occupants than the higher, thus minimizing the total energy. There are more spins aligned with the field than opposing it. These surplus microscopic magnets constitute, when summed, the macroscopic magnetization, which is of course along the field, or in the usual jargon, it is longitudinal. Thus, in equilibrium, before perturbation there is a longitudinal magnetization. However, this longitudinal magnetization is very difficult to measure; we need a trick to extract information from the spin system. We must therefore disturb the system, disrupt its balance, and see how it reacts. The difference between the two energy levels is determined by B_0 and the gyromagnetic ratio. If we want our spins in equilibrium to react to an external excitation, we must influence them with an external source that has an energy exactly equal to this difference. This unequivocally determines the energy. Since, as Einstein taught us, energy and frequency are also proportional ($\Delta E = hf$), we can now understand why it is necessary to impose radio waves with a well-defined frequency to perform NMR.

As already described, relaxation immediately follows excitation. This relaxation is determined by the interactions of the nuclear spins with their environment and with other spins. Two separate relaxation phenomena can be distinguished, corresponding to different thermodynamical processes. One is

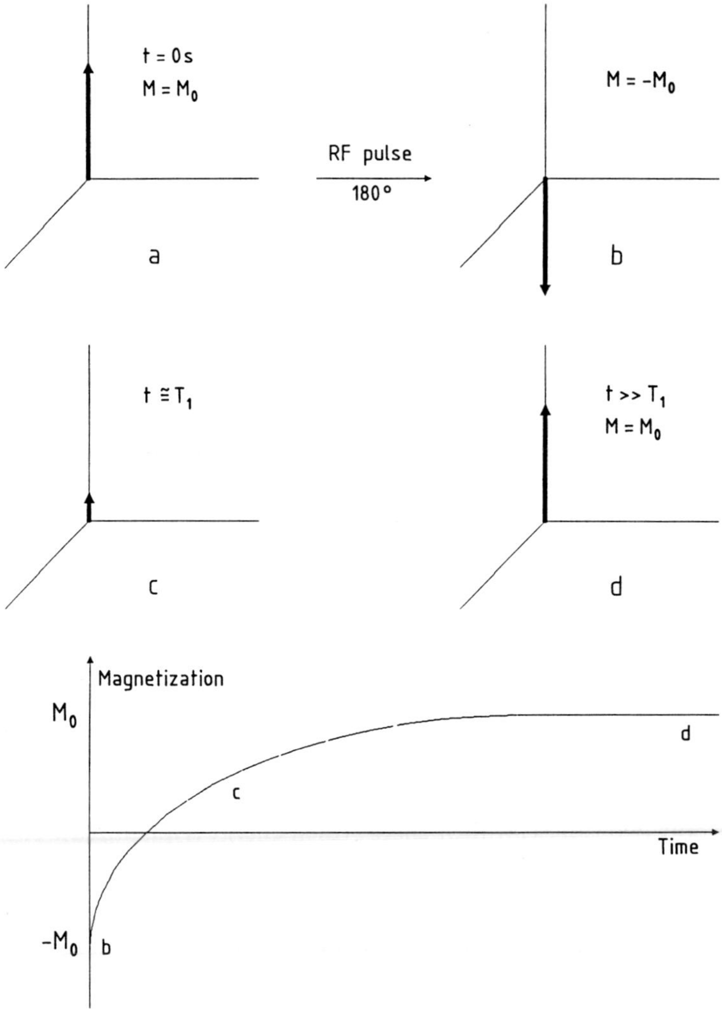

Fig. 1.3. If the spins are not disturbed they are in an equilibrium arrangement, adding up to create the longitudinal magnetization M_0. A so-called 180-degree pulse forces the macroscopic magnetization to point in the opposite direction. This then gradually returns to its original direction, governed by an exponential law with a specific time constant T_1. After a time much longer than this T_1, the magnetization returns to its equilibrium state

called longitudinal relaxation. This is the mechanism that the system uses to get rid of the extra amount of energy which it has received from the external source. It consists in rebuilding the original longitudinal magnetization by restoring the original distribution of the spins over the two energy levels. This restoration is performed with a characteristic time, called T_1. This process is governed by the way in which the spins and their environment exchange energy. Since this is obviously dependent on the environment, spins in fluid, for example, have a

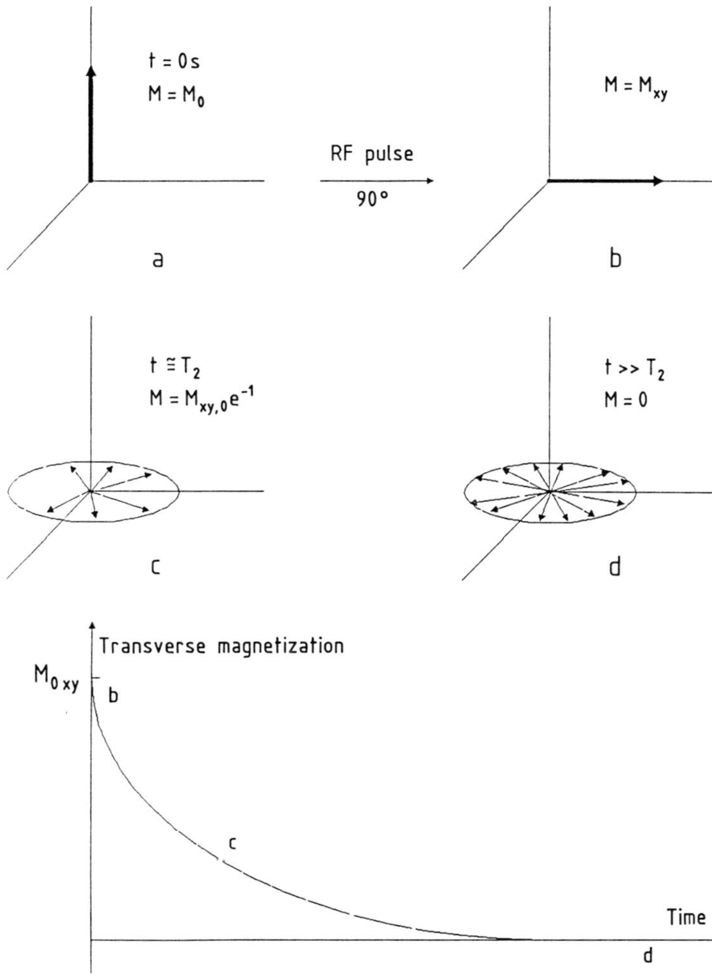

Fig. 1.4. Another excitation pulse tilts the magnetization over an angle of 90 degrees into the plane perpendicular to the magnetic field. Nuclear magnets in this situation are found to perform a rolling motion (precession) resembling that of a top. In practice, different spins have slightly different precession frequencies, causing a fanning out of their orientations and a gradual reduction of the resulting transverse magnetization

different T_1 from spins in bone. The other mechanism is transverse relaxation. Here the energy is not affected; it simply causes a loss of order. The spins, which are aligned just after an excitation pulse, gradually fan out and lose any coherence. The signal thus dies out with a time characteristic to this process and different from the previous one, called T_2. As a very general rule one can say that spins in a fluid environment lose their coherence slower than those in a more rigid environment. Thus, for instance, spins in CSF have a higher T_2 than spins in bone.

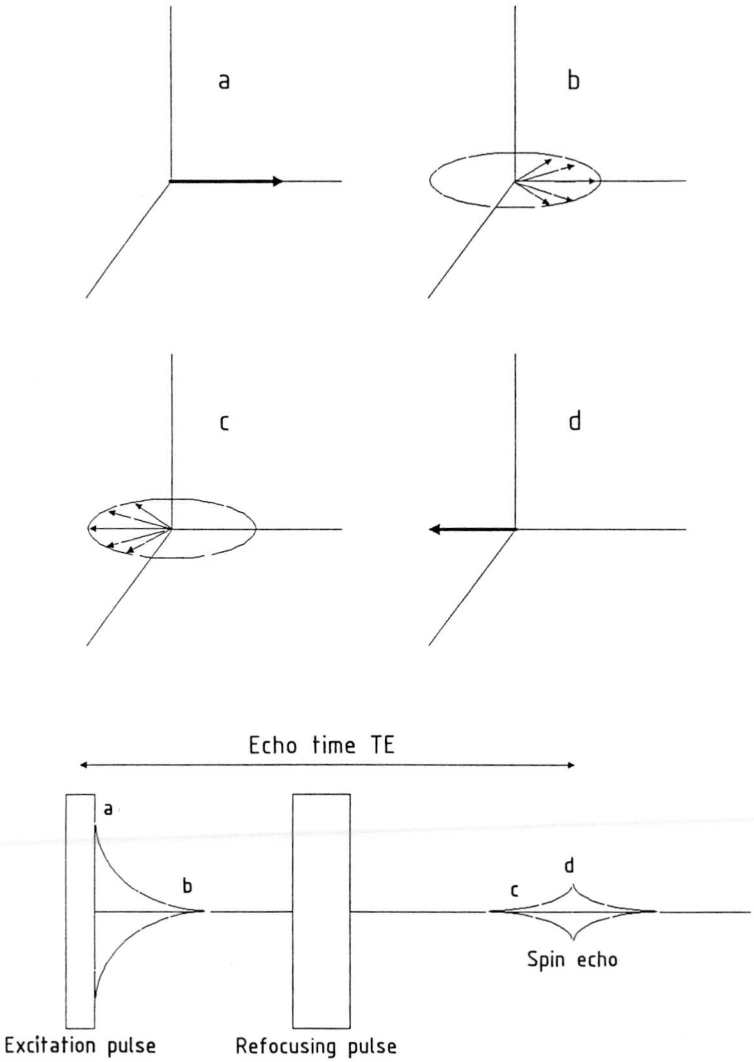

Fig. 1.5. The harmony lost due to the spin-spin interactions can be restored with the help of a second pulse, a 180-degree pulse. At time *a* the spins are aligned, just after a 90-degree pulse. They immediately start to lose their unanimity and fan out. At time *b* the resultant signal has therefore decreased. At time *c* a 180-degree pulse is given. Spins are flipped over to the other side. Now their movement results in realignment of the spins. At time *d*, twice the time between the first and the second pulse, a new signal, the spin echo, appears. The total time is appropriately called the echo time, abbreviated TE

In order to take advantage of this second parameter, the signal has to be collected in a peculiar way. Instead of measuring immediately after the excitation, one waits for the order to be lost. All the spins are fanned out and the total signal is zero. The trick then consists in applying a second excitation pulse, commonly called the 180-degree pulse, which forces the spins to reorder.

All this results in a new signal, called the spin echo. It arises at a time after the refocusing second pulse which is exactly the same as the time between the two first pulses. The total time between the first pulse and the spin echo is called the echo time, or in abbreviation TE. This echo time is one of the main parameters in imaging. It determines the signal and contrast of the resulting image to a large extent. Tissues with a long T_2 will show bright in an image with a relatively long TE, because T_2 is a measure of the tissue spins' ability to reorder and thus create signal.

1.2 Looking at Metabolism Using NMR

1.2.1 Chemical Shift

Molecules consist of a number of atoms bonded together via electrons. These electrons are also influenced by the magnetic fields applied during an NMR experiment, and therefore they react to this external field by shielding the atoms which they surround. This results in different magnetic fields experienced by the different atoms in a molecule. The Larmor formula must be adapted to include this phenomenon, called chemical shift:

$$\omega_0 = \gamma B_0 (1 - \sigma)$$

The chemical shift depends on the position of the atom in the molecule and on the chemical environment; atoms in the same molecule thus have a different resonance frequency according to their position and their chemical environment. This provides us with a trick to differentiate between nuclei, for example, between protons in water and protons in fat, or phosphorus in phosphocreatine and inorganic phosphate.

The information that we gain from a spectroscopic experiment not only tells how many NMR-different species there are in a sample but also provides us with a measurement of their relative concentrations. In some cases other information is also available (e.g., pH). We go deeper into details in the next section.

1.2.2 How and Where to Use Spectroscopy?

Biomedical applications of NMR spectroscopy remain restricted mainly to three nuclei: hydrogen (^1H), carbon (^{13}C), and phosphorus (^{31}P). There is a potential use of lithium (^{7}Li), fluorine (^{19}F), sodium (^{23}Na), potassium (^{39}K), and some other exotic nuclei, but we will not go into these applications in this text, which concentrates on techniques in sports medicine. The main criteria for selecting these nuclei for NMR are the following. (a) The γ value should be high enough, otherwise the intrinsic sensitivity is insufficient. (b) The metabolites to be studied must be present in a concentration in the order of millimoles. (c) The molecule (or a part of it) should be mobile, otherwise the signal is quenched.

^{1}H Spectroscopy. The big advantage, and at the same time the biggest problem, for proton spectroscopy is the fact that protons are found in almost every biochemical substance, and especially in water (H_2O) and fat (mainly $-CH_2-$ chains). This is the reason why, for imaging, protons are used to collect the data necessary to reconstruct images. Due to their abundance and sensitivity they deliver an enormous signal compared to other nuclei. It is this enormous signal that causes problems in spectroscopy, where most of the time one is interested in signals arising from substances such as lactate, choline, creatine, and so on. It is clear that the signal of the latter is much weaker, and we must therefore use special techniques to avoid the drowning of the weak signal by the about 10000 times stronger signal of water and fat. Several techniques have been proposed to achieve this. Most of them destroy rather efficiently the unwanted signal, but at the same time they produce some inherent deformation of the spectrum. Also, these editing techniques of course complicate the experiment. However, if one succeeds in overcoming this problem, one has direct access to very interesting biochemical substances. The most important one in the context of sports medicine is lactate. The importance of lactate in muscle metabolism (and also cancer research) is beyond doubt. The signal of the methyl group of lactate however is obscured by the signals from the fat. Nevertheless, its interesting spectroscopic properties make it a favourable candidate for detection, providing the appropriate editing technique is used.

^{13}C Spectroscopy. Just as in the case of protons, one can find carbon atoms in almost every biochemically important substance, and using carbon NMR spectroscopy would thus seem ideal for our purposes. However, there are some serious drawbacks. First of all, of the three isotopes of carbon (^{12}C, ^{13}C, and ^{14}C) only ^{13}C is susceptible to NMR. Since the abundance of this isotope is only 1.1%, only a small quantity of carbon atoms is available; this of course reduces the overall sensitivity. Secondly, the γ value of carbon is relatively low, decreasing once more the signal amplitude considerably. Thirdly, since in most molecules the carbon atoms are coupled very strongly with protons, the spectra tend to be very complicated. In order to reduce this complexity another radio frequency field must be applied to annihilate the influence of the protons. In short, carbon spectroscopy requires a very sensitive technique to gather the information needed. Therefore one should work at a field as high as possible and average many experiments to obtain a reasonable signal. Fortunately, the second decoupling field can cause an increase in signal if applied at the right times during the experiment. This nuclear Overhauser enhancement can

theoretically increase the signal by a factor of 3, in the best of worlds. Some other very ingenious techniques have been proposed to transfer the magnetization of the protons to carbon atoms, taking advantage of the better sensitivity of the latter. Experiments that can be performed in vivo on humans include the determination of the ratio of saturated/unsaturated fats, detection of glycogen, and dynamic studies using ^{13}C-labeled compounds.

^{31}P Spectroscopy. Phosphorus offers the opportunity to study the energy metabolism in a very direct and uncomplicated way. In fact, the in vivo spectrum consists primarily of seven fairly well-delineated peaks. Three of these are attributed to the three phosphates in an ATP molecule, one arises from phosphocreatine, one from inorganic phosphates, and the other two stem, respectively, from the phosphomonoesters and sugar phosphates and from the phosphodiesters. The only drawback here is to low sensitivity of ^{31}P, resulting in the necessity to perform several averaging acquisitions per experiment. A typical phosphorus experiment consists of imposing a work load for the muscle under study and following the metabolism. Not only the relative concentration of the above-mentioned metabolites can be monitored but also the internal pH. In fact, the position of the inorganic phosphate (Pi) peak shifts with the pH. This can be easily understood: the Pi peak is the weighted average of the peaks arising from $H_2PO_4^-$ and HPO_4^{2-}, which interreact according to the formula:

$$H_2PO_4^- + H_2O = HPO_4^{2-} + H_3O^+$$

The equilibrium of the reaction is, of course, influenced by the concentration of hydroxonium ions, or in other words the pH. Thus, the higher the pH, the more the equilibrium is shifted towards HPO_4^{2-}, and consequently the more the peak drifts in the direction of the HPO_4^{2-} peak. This phenomenon can be exploited quantitatively through the Hasselbach-Henderson relation:

$$\text{pH} = \text{pK} + \log(\text{Pi shift} - 3.29) - \log(5.68 - \text{Pi shift})$$

where Pi shift is the difference between the Pi peak and the phosphocreatine peak.

1.3 Making an Image with NMR

1.3.1 The Nature of NMR Images

Digital Images. To understand how images can be made using NMR it is necessary to have some basic idea of the nature of digital images. Figure 1.6 shows how a two-dimensional collection (array) of numbers can be translated into an image, by simply relating the position of each number in the array to a spatial position within the image and interpreting the value of each number as a visual intensity level. The individual image elements are called pixels, and their

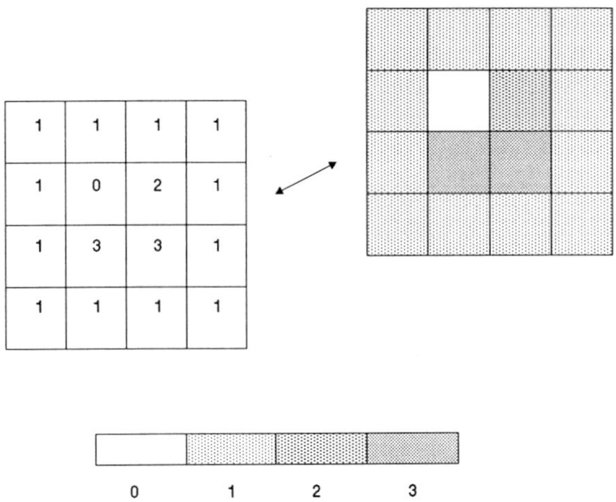

Fig. 1.6. A digital image, its numerical counterpart and the gray scale used for translating one into the other

size determines the spatial resolution of the image. The numerical values of the different elements can be made to correspond to a physical quantity, such as X-ray attenuation in computed tomography or temperature in thermography. The precision with which these values are determined (measurement process) and numerically represented (digitization process) determines the contrast resolution of the image.

NMR Images. The physical quantity underlying an NMR image is the local intensity of the radiowaves emitted by the hydrogen nuclei in the body after a suitable preparation. This preparation includes radio frequency excitation pulses, evolution of the nuclei in the presence of magnetic field gradients, and various time delays. In general, in order to allow time for these preparatory steps, the signals observed are of the spin-echo type. The preparatory steps are necessary for localizing the origin of the signals concerned as well as for optimizing the differences between the signals emitted by tissues of interest, i.e., optimizing contrast.

1.3.2 Contrast

Relaxation. As explained above, NMR is based on the very special magnetic properties of some nuclei. As a consequence of these properties, the nuclei, when placed in a magnetic field, act together to produce a macroscopic magnetization, pointing along the direction of the field. In this situation, the nuclei can absorb and reemit radiowaves of a specific frequency (the so-called resonance frequency). Using radiowave excitation, the magnetization can be forced away from its equilibrium orientation and tipped over a certain angle (tip angle). Once tipped away from equilibrium, the magnetization carries out a precession (i.e., a rolling motion similar to that of a spinning top) around the

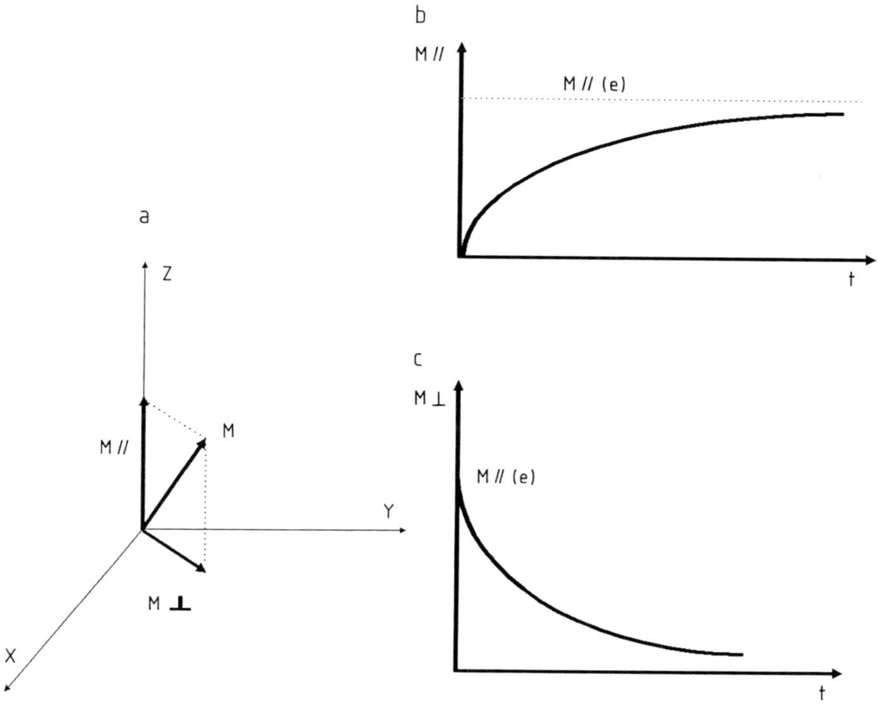

Fig. 1.7. **a** Decomposition of the magnetization vector into its longitudinal and transverse components. Recovery of the longitudinal (**b**) and transverse (**c**) components after a 90° pulse

direction of the field. The precession frequency is exactly the same as the resonance frequency of the nuclei involved. Moreover, this frequency is found to be proportional to the strength of the external magnetic field. Upon termination of radiowave excitation, the nuclear spins tend to return to equilibrium, with their collective magnetization pointing along the field. The return to equilibrium is called relaxation, and the processes that make it possible are called relaxation processes. With respect to relaxation, the macroscopic magnetization is best seen as the sum of a longitudinal component (pointing along the field) and a transverse component (perpendicular to the direction of the field) (Fig. 1.7a). The two components are found to relax quite differently.

T_1 *Relaxation.* Longitudinal magnetization returns to its equilibrium value $M /\!/ (\mathrm{e})$ following an exponentially increasing function. After flipping the magnetization into the transverse plane, for instance, recovery of the longitudinal component can be described by:

$$M /\!/ (t) = M /\!/ (\mathrm{e})\,[1 - \exp(-t/T_1)]$$

(Fig. 1.7b). The constant T_1 is the so-called spin-lattice relaxation time (see Sect. 1.1.2).

T_2 Relaxation. Transverse magnetization is zero in equilibrium, and this situation is reached following an exponential decay:

$$M\perp(t) = M\perp(0)\exp[-t/T_2],$$

where the constant T_2 is the spin-spin relaxation time. Of course, for a flip into the transverse plane (Fig. 1.7c), $M\perp(0) = M\parallel(e)$.

Echoes and Their Contrast Mechanisms. Contrast, i.e., differences in image gray levels representing the various features, originates in NMR from differences in the radio frequency signals from the tissues involved. From the discussion in previous paragraphs, it is clear that the relative intensity of the echo signals from these tissues is determined by their intrinsic parameters proton density, T_1, and T_2 and the extrinsic or instrumental parameters TR and TE. The mechanisms determining T_1 and T_2 values in different tissues are reviewed in the literature and are only summarized here. The T_1 value characterizing a specific tissue sample reflects how efficient the magnetic energy transfer between the hydrogen nuclei and their environment is. This in turn determined by the mobility of the nuclei in their molecular environment. Solids and pure liquids have very long T_1, fatty tissue is characterized by very short T_1, while protein solutions have intermediate T_1. On the other hand, T_2 relaxation processes include energy-exchanging events as well as the loss of phase coherence of individual spin precessions (fanning out) described in Sect. 1.1.2. As a consequence, T_2 tends to be much shorter than T_1 in biological tissue. Differences in T_2 are again related to the mobility of the nuclei: T_2 values are very long for pure liquids and decline as the nuclei become less mobile (protein solutions, solids). The instrumental factors TR and TE can be used to influence the relative importance of the intrinsic parameters in determining image contrast. As can be inferred from the above discussion, T_1-weighted images are obtained using short TR and short TE, T_2-weighted images need long TR and short TE, proton density weighted images correspond to long TR and long TE.

1.3.3 Localization

Fields and Frequencies. The radiowaves that are reemitted by the nuclear spins contain information on the local environment in which the nuclei evolve and can therefore generate image contrast through the mechanisms outlined in Sect. 1.3.2. In order to organize this information in image form, however, the location from which these radiowaves originate must also be determined. Since the resonance frequency is proportional to the field strength, this can be achieved by making the field strength position dependent: field strength and, hence, resonance frequency can then be used as a label corresponding to position.

Field Gradients. In practice, use is made of linear field gradients along the axes of the reference system tied to the magnet. The magnetic field strength is made

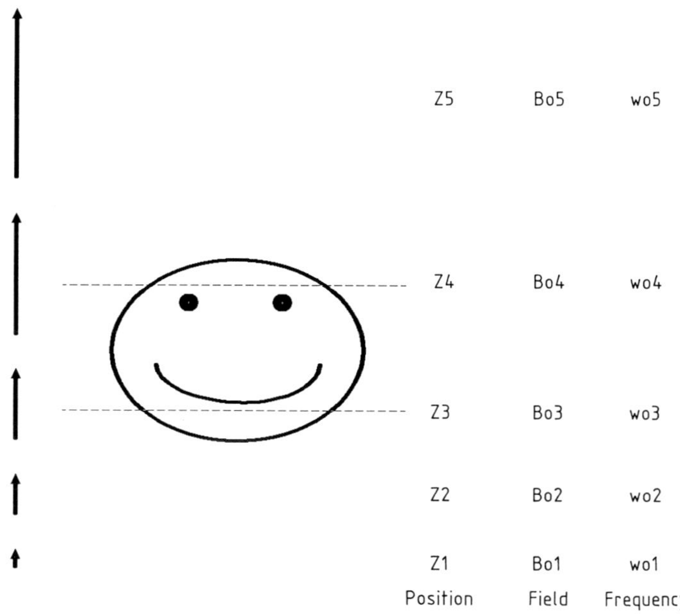

Fig. 1.8. Selective excitation. Radiowaves contain only frequency components $\omega(z_3) < \omega < \omega(z_4)$

to vary linearly along these axes, for example, x, following:

$$\frac{dB}{dx} = \text{constant} = G_x$$

The gradient strength G_x is the constant rate of change of the magnetic field strength as one walks along the x-axis. In the presence of this field gradient, resonance frequencies are position-labeled by:

$$\omega(x) = \gamma(B_0 + G_x x)$$

where ω designates the angular resonance frequency, x the x-coordinate, v the gyromagnetic ratio of the nuclei, and B_0 the strength of the static magnetic field in the absence of gradients.

Three Steps Towards Localizing Signals. The first of the three coordinates is determined by the selection of the slice that is to be imaged. This slice is selected by irradiating the sample with a radiowave pulse having a very precise frequency content, in conjunction with a single field gradient, for example, along z (Fig. 1.8). Only the nuclei with a resonance frequency present in the pulse are excited, i.e., only nuclei with the right z-coordinates contribute to the image.

Spatial encoding within the slice selected in the previous step is achieved using the two remaining gradients, along x and y. In one of these directions,

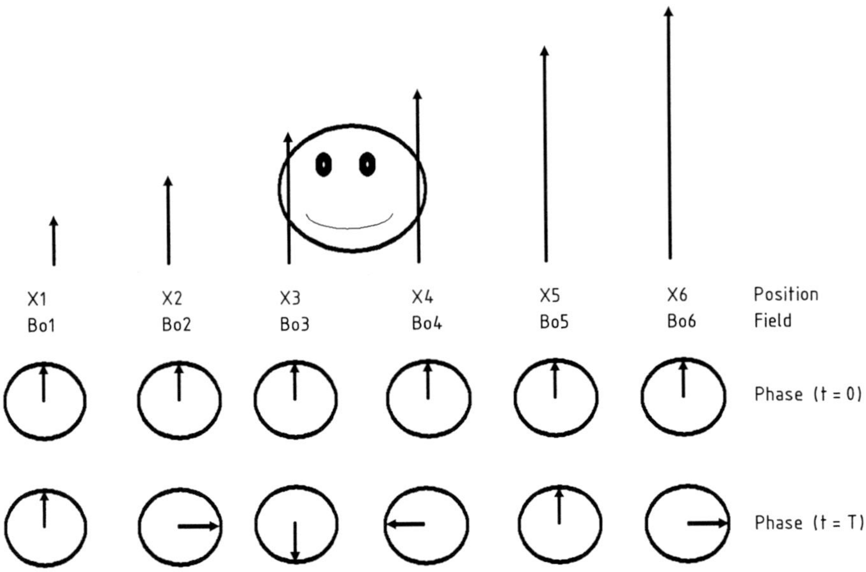

Fig. 1.9. Phase encoding of position in the image plane. Nuclei in different positions along the phase-encoding gradient gain different phases in the same period of time

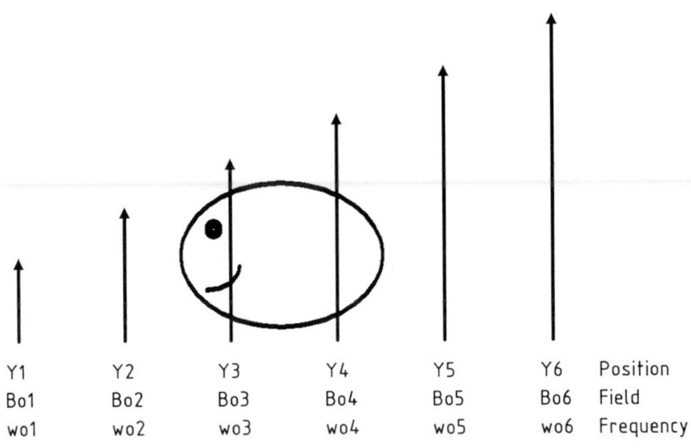

Fig. 1.10. Frequency encoding of position in the image plane. Fourier analysis allows separation of the radio signals, emitted at $\omega(x_3)$ and $\omega(x_4)$, for instance

spatial information is encoded as phase differences in the precessional motion of the individual nuclei at different locations and therefore as phase differences in the signals they emit. A gradient along this direction (say x) is switched on during a certain amount of time (Fig. 1.9). Whereas before the gradient ($t = 0$), spins in different positions along the x-axis precess synchronously, the presence of the gradient during time T forces some nuclei to precess faster than others,

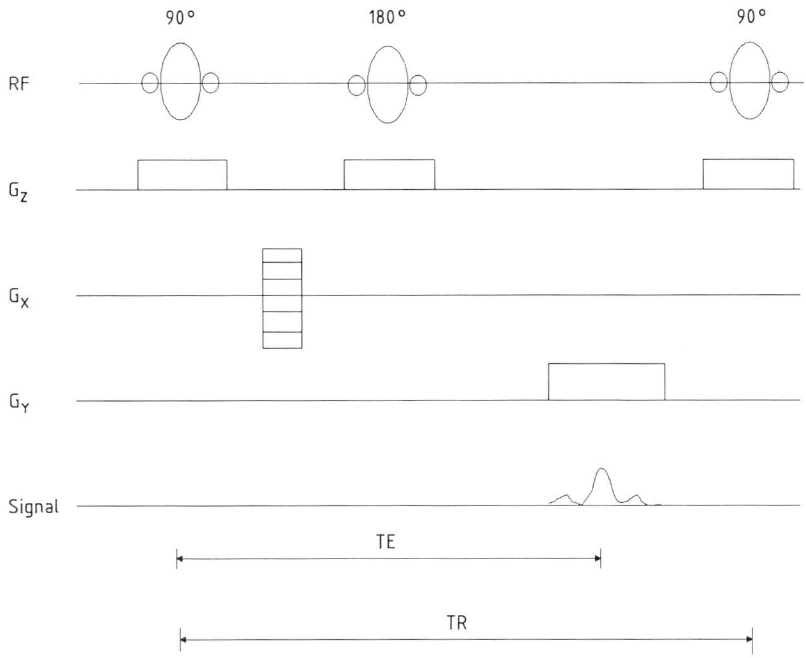

Fig. 1.11. A spin-echo imaging sequence with selective excitation along z, phase encoding along x, and frequency encoding along y

thus introducing phase differences and therefore determining the relative weight of the contributions from nuclei at different positions in any signal collected after the gradient is switched off. Since the same phase can be obtained after one or more precessional motions, it is impossible to label positions uniquely by phase. This problem is overcome by repeating the same measurements a number of times, each time with a slightly different phase-encoding gradient.

The third dimension (in our example the y coordinate, Fig. 1.10) is encoded by applying a gradient while the emitted radiowaves are sampled. Again, nuclei at different y positions have different resonance frequencies, and a simple Fourier transform suffices to extract the frequency components and their position along the read gradient.

Putting It Together. Figure 1.11 shows how these different tricks are combined to make an image. A spin-echo signal is acquired in the presence of a read-encoding gradient, providing resolution along y. Since a gradient along z is present during the two exciting pulses, and since both pulses are selective, only hydrogen nuclei in a well-determined slice perpendicular to the z-axis contribute to the signal. The signal itself consists of contributions located along the x-axis and having different phases, as determined by the value of the phase-encoding gradient. In order to identify these contributions, the basic measuring cycle is repeated, this gradient being stepped through as many values as there are pixels along the x-axis.

From the spin-echo signals obtained in this way, the image can be obtained by Fourier transformation, a mathematical technique available on computer systems used for NMR. The technique decomposes the signals measured in the time domain into their frequency components, hence into their contributions originating from different locations. As shown in Sect. 1.3.2, image contrast is determined by the duration TR of the basic measuring cycle and the echo time TE. These times function as the main instrumental parameters and allow image contrast to be weighted according to any of the intrinsic tissue characteristics:

Proton density weighted: long TR, short TE
T_1 weighted: intermediate TR, short TE
T_2 weighted: long TR, long TE

The total duration of the image acquisition is given by:

$$T_m = \mathrm{TR}\, n\, r,$$

where n is the image matrix size (e.g., 256) and r is the number of times the same signal is measured for signal averaging purposes (for noise reduction; measuring the signal r times increases the signal-to-noise ratio by $r^{1/2}$).

Typically, a T_1-weighted image takes:

$$T_m = 0.5\,\mathrm{s} \times 256 \times 1$$
$$\sim 2\ \mathrm{min}$$

or more. Proton density and T_2-weighted images take:

$$T_m = 2\,\mathrm{s} \times 256 \times 1$$
$$\sim 8\ \mathrm{min}$$

or more. These relatively long imaging times are alleviated by the fact that multislice acquisitions are possible; while waiting for each slice to recover from one measuring cycle, information on other slices can be acquired.

1.3.4 Special Imaging Techniques

Fast Imaging. As shown in the previous section, the standard spin-echo imaging using a 90° tip angle and a 180° pulse for refocusing the spins, measuring time can be calculated by multiplying the number of phase-encoding steps by the repetition time TR and the number of acquisitions that are performed for signal averaging. The repetition time is dictated by the rate at which the hydrogen nuclei recover after the 90° flip (i.e., by T_1) and the type of contrast wanted. This means that for a typical examination, which includes several orientations and/or several contrast options, an average duration of 45 min is needed. From the point of view of patient comfort, avoidance of motion artifacts, and clinical efficiency, it is clearly desirable to speed up the imaging process. To do this in conventional spin-echo imaging is very difficult.

Fig. 1.12. Longitudinal and transverse components of a magnetization vector tipped over a small angle

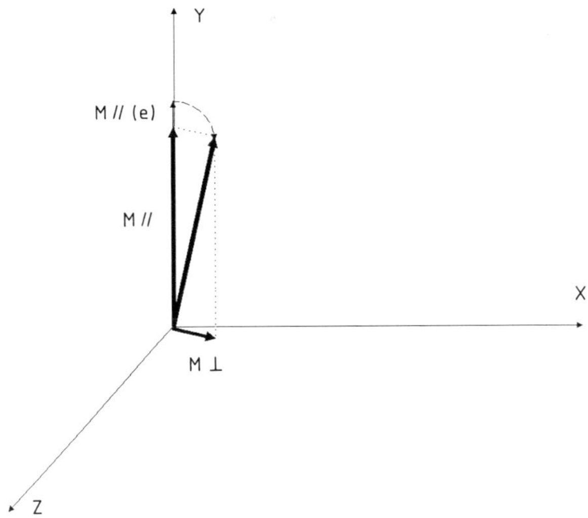

Shortening TR affects contrast behavior and signal-to-noise ratio. Reducing the number of phase-encoding steps has an immediate effect on spatial resolution or field of view. The number of acquisitions is at least 1, and multiple acquisitions may be needed for improving the signal-to-noise ratio.

Solutions to this problem have been proposed in the past. These vary from techniques that use additional radiowaves to force the return to equilibrium of the protons, to techniques that actually reduce radiowave input, using small tip angles. FLASH (fast low-angle shot) and FISP (fast imaging with steady precession) are two typical representatives of the latter solution, which until now has been the only one used extensively in the clinic. In the following, we restrict our discussion to these two sequences.

As shown in Fig. 1.12, using a small tip angle means that most of the magnetization remains longitudinal at all times, and that a relatively large gain in transverse magnetization is paid for by a relatively small loss of longitudinal magnetization. Consequences of this strategy are that recovery takes less time as the tip angle is decreased, and that we can no longer use Hahn echoes to reverse the dephasing action of field inhomogeneity. Since a 180° pulse would invert the (large) longitudinal magnetization, it would take even longer than with conventional techniques to recover. Gradient echoes, however, can still be used. They allow very short TE values and reduced radio frequency power, but they rephase only the gradient-induced dephasing contributions, rendering images sensitive to other field inhomogeneities.

Figure 1.13 shows that both FLASH and FISP sequences work. Note that the two sequences are identical until image acquisition has taken place. Then, the transverse magnetization is destroyed by a strong spoiler gradient in the case of FLASH, while it is optimized as far as possible by the end of the basic cycle in the case of FISP. The consequence of this is that FLASH images are essentially T_1 weighted (longitudinal component predominant), while FISP

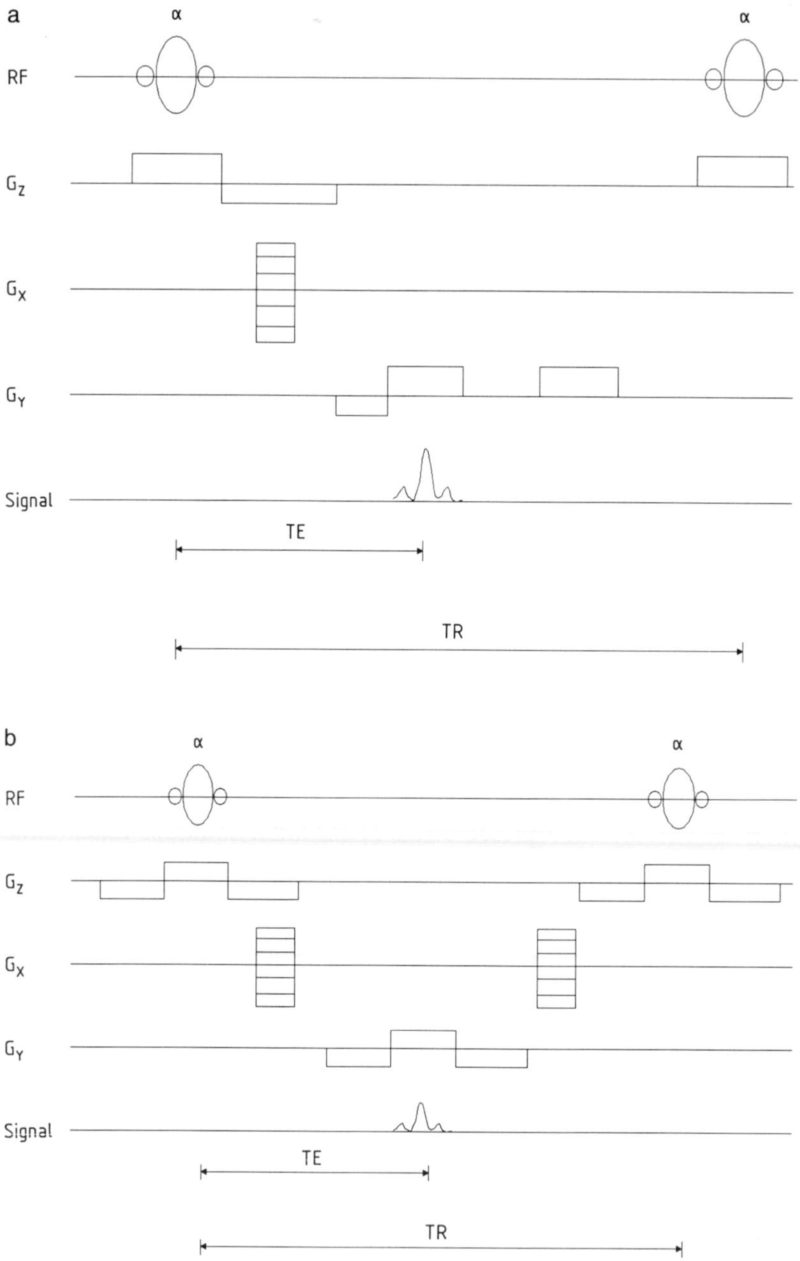

Fig. 1.13. **a** FLASH basic cycle. The strong gradient at the end of the cycle effectively spoils residual transverse magnetization. **b** FISP basic cycle. The symmetrical structure of gradients before and after the measuring interval optimizes transverse magnetization at the end of the cycle

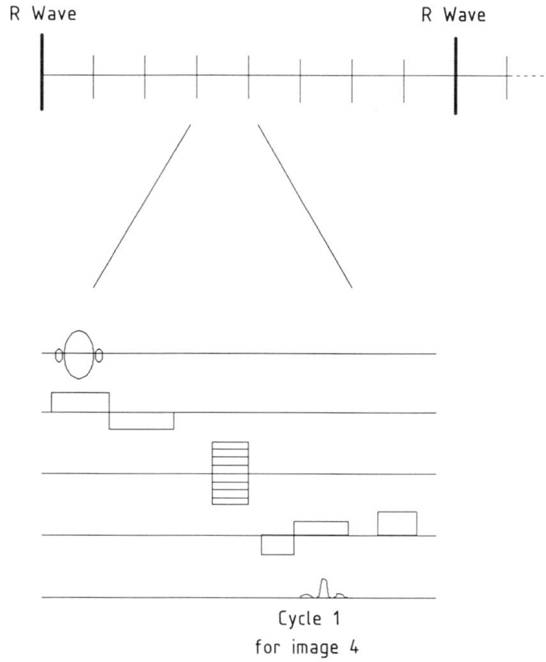

Fig. 1.14. ECG-triggered measuring scheme for obtaining a dynamic series of NMR images (single slice) covering one heart cycle

can be used for T_2-weighted imaging (effect of residual transverse magnetization).

In comparison to conventional spin-echo imaging, in both sequences short TR can be combined with reasonable signal-to-noise ratio.

Three-Dimensional Applications. Although FLASH and FISP allow an image to be made in one breathhold ($T_m = 256 \times 0.04\,\text{s} \times 1 = 10\,\text{s}$), their short cycle time allows other, more interesting applications. One of these is three-dimensional imaging. Instead of imaging slice by slice using selective excitation, it is possible to collect information over the entire volume of interest, introducing spatial resolution in the slice direction by an additional phase-encoding gradient, which must be stepped through as many values as there are pixels in this direction. The total imaging time is now:

$$T_m = \text{TR}\, m\, n\, r$$

(m is the number of steps of this phase-encoding gradient). This can be kept reasonably short, typically:

$$\begin{aligned} T_m &= 0.03\,\text{s} \times 256 \times 128 \times 1 \\ &\sim 16\,\text{min} \end{aligned}$$

After this time, we have a collection of 256 × 256 × 128 voxels (i.e., three-dimensional equivalent of pixels) covering the entire volume of interest (e.g., head, knee). Specialized imaging workstations exist that allow sections through this volume to be reconstructed in any desired orientation. Due to the near-isotropic spatial resolution of the original information, these postreconstructed images are of excellent quality.

Dynamic Imaging. The short cycle time of FLASH and FISP can also be used for dynamic NMR applications. As shown in Fig. 1.14, the RR interval of 900 ms of a typical human heart can accomodate about 30 measuring cycles of 30 ms each. ECG triggering allows this procedure to be repeated for the required number of phase-encoding steps, so that gradually images corresponding to 30 different time intervals covering the heart cycle can be collected. Imaging time can be calculated as:

$$T_m = \mathrm{RR}\, n\, r$$

For heart studies, $n = 128$ and $r = 4$ typically, and it therefore takes about:

$$\begin{aligned} T_m &= 0.9\,\mathrm{s} \times 4 \times 128 \\ &\sim 8\ \mathrm{min} \end{aligned}$$

to obtain a complete dynamic series of images covering the heart cycle. Again, a specialized image workstation facilitates display and interpretation of these images.

MR Imaging in Sport Injuries to the Osteoarticular System

2 The Knee Joint

M. Shahabpour, M. Osteaux, and P.-P. Casteleyn

2.1 Introduction

The most commonly injured joint is the knee due to the tremendous stresses applied on its structures in athletic activities. Knee injuries are also the leading cause of long-term disability of athletes in a broad spectrum of sports (Larson 1985; Kujala 1986; Kaye 1989b). Knee injuries may develop from overuse syndromes in running or swimming, from deceleration injuries in basketball, sprint or soccer, or from violent, direct blows in football, hockey, rugby, or skiing (Kennedy 1979; Hughston and Barrett 1983). Knee injuries are also related, less expectedly, to such activities as middle-distance running and javelin throwing (King 1986b).

Conventional radiographic examination remains a mandatory tool. A study of bone and soft-tissue components is necessary in the investigation of any joint pathology: evaluation of bone mineralization, detection of intraarticular or intraosseous calcifying processes (mineral opacities), study of the mechanical axes, and dynamic views. Comparison with the contralateral joint is a useful complement (Resnick 1989; Malghem and Maldague 1989; Sintzoff et al. 1989).

Arthrography is a nonexpensive diagnostic modality but needs intraarticular injection of contrast material and is dependent upon the operator and his experience. In addition, the accuracy rates remain low in the lateral knee compartment (Levinsohn and Baker 1980; Resnick 1989; Sintzoff et al. 1989).

Diagnostic arthroscopy is a more invasive method that requires local or general anesthesia. Its accuracy is also operator dependent and requires experience to attain ideal value for intraarticular pathology. Arthroscopy must be considered mainly in combination with therapeutic procedures (Levinsohn 1980; Sherman and Fox 1986).

During recent years *magnetic resonance imaging technology* (MRI) has shown an explosive expansion: development of dedicated coils specifically designed for the limbs, improvement of pulse sequences, advances in computer software, particularly the use of fast imaging and three-dimensional acquisitions, and development of specialized image processors in image reconstruction (Mink 1987; Shahabpour 1989a).

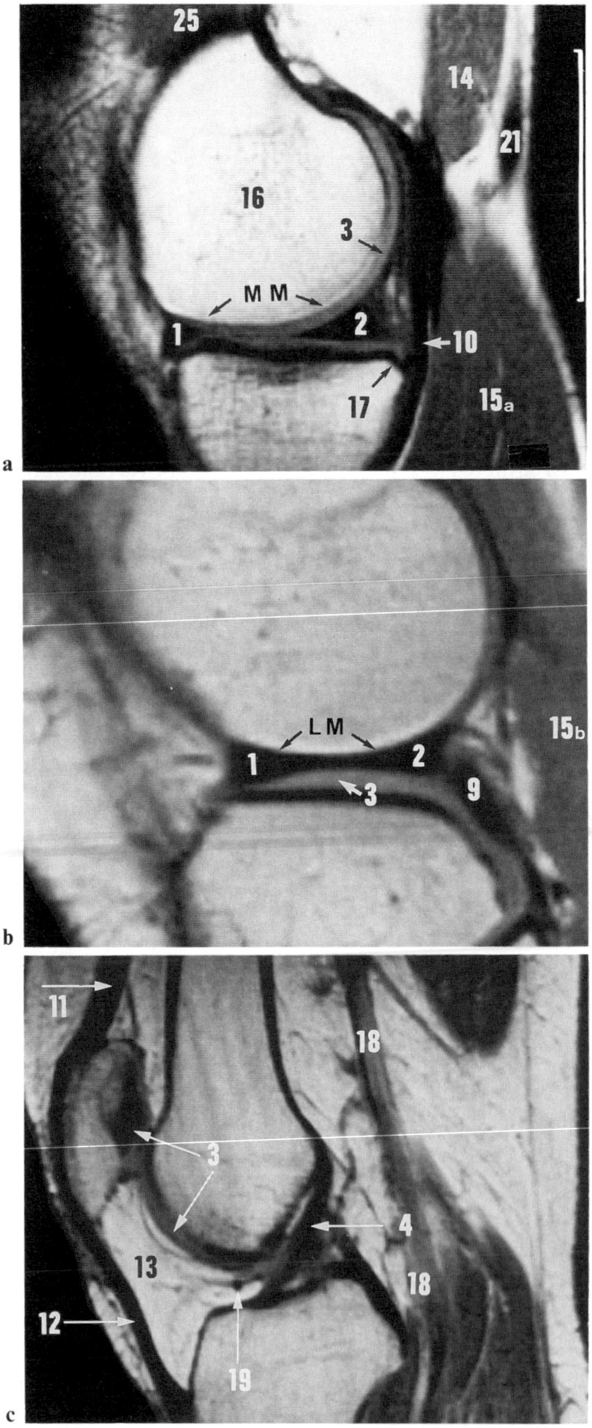

Fig. 2.1 a–c

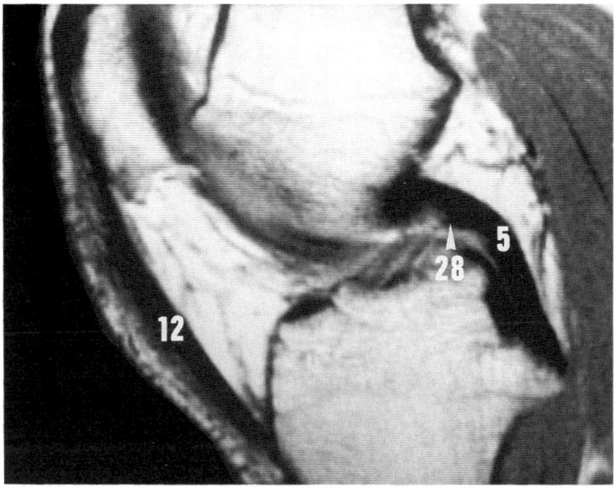

d

Fig. 2.1 a–d. Sagittal anatomy. **a** Section through the medial meniscus (*MM*; SE 600/15). **b** Section through the lateral meniscus (*LM*; SE 2200/22). Note: (a) the bow-tie appearance of the peripheral portion of the meniscus; (b) the thickened appearance of the tibial cartilage due to chemical shift artifact. **c** Section through the intercondylar area at the level of the anterior cruciate ligament (SE 2200/22). **d** Section through the intercondylar area at the level of the posterior cruciate ligament (SE 2200/22). *1*, Anterior meniscus horn; *2*, posterior meniscus horn; *3*, hyaline articular cartilage; *4*, anterior cruciate ligament; *5*, posterior cruciate ligament; *9*, popliteus tendon; *10*, posterior articular capsule; *11*, quadriceps tendon; *12*, patellar tendon; *13*, infrapatellar fat pad; *14*, semimembranosus muscle/tendon; *15a*, gastrocnemius muscle, medial head; *15b*, gastrocnemius muscle, lateral head; *16*, medullary bone; *17*, cortical bone; *18*, popliteal vessels; *19*, transverse geniculate ligament; *21*, semitendinosus muscle/tendon; *25*, vastus medialis; *28*, Humphrey ligament

2.2 Multiplanar Anatomy

Views of the sagittal, axial, and coronal anatomy of the knee are presented in Figs. 2.1–2.3. The signal intensity of the musculoskeletal structures depends on imaging parameters such as proton density, longitudinal and transverse relaxation times T_1 and T_2, and on operational parameters such as TE and TR in spin-echo imaging, in addition to flip angle in gradient-echo imaging. Except for synovial fluid and hyaline cartilage, spin-echo gray scale values of the musculoskeletal components do not differ much from one pulse sequence to another. The net contrast between the musculoskeletal structures results from differences in the mobile proton content of adjacent anatomical tissues.

Cortical bone gives a very low signal due to its poor content of mobile protons and appears as a dark line surrounding the cancellous bone. Due to the large proportion of fat, which has a high content of mobile protons, *medullary bone* has a high signal intensity on spin-echo images. On gradient-echo images, its signal is lower.

Hyaline articular cartilage shows an intermediate value on the gray scale with spin-echo pulse sequences T_1 and proton density. It can be clearly

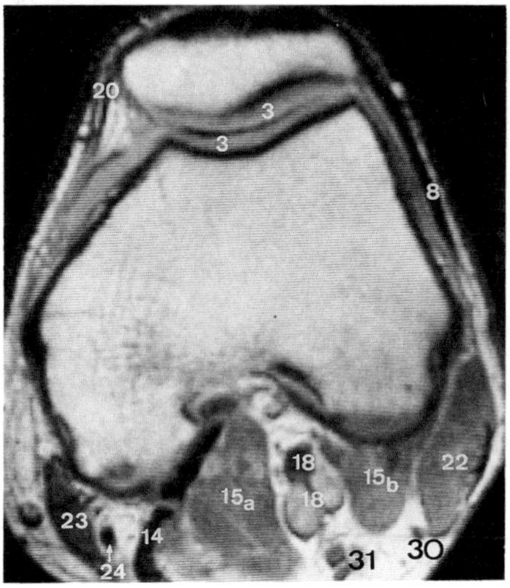

Fig. 2.2. Axial anatomy. Section through the patellofemoral joint (SE 2200/22). *3*, Hyaline articular cartilage; *8*, iliotibial band; *14*, semimembranosus muscle/tendon; *15 a*, medial head of gastrocnemius muscle; *15 b*, lateral head of gastrocnemius muscle; *18*, popliteal vessels; *20*, medial patellar retinaculum; *22*, biceps femoris muscle/tendon; *23*, sartorius muscle; *24*, gracilis tendon; *30*, common peroneal nerve; *31*, tibial nerve

distinguished from meniscal fibrocartilage, which appears dark. The difference could be explained by the water content of their collagen fibers. Type II collagen fibers of the hyaline cartilage contain more hydroxylysine than type I collagen fibers present in menisci and tendons. The signal intensity of the articular cartilage decreases on T_2 SE pulse sequences and increases on GE pulse sequences.

The *menisci* can be clearly visualized on coronal and sagittal planes. MRI has the ability to visualize intramural (intrasubstance) changes within the menisci.

Ligaments, tendons, and capsule appear as dark structures, contrasting markedly to surrounding fat or muscles. The *posterior cruciate ligament* (PCL) can be seen as a thick dark arc on mediosagittal views. The *anterior cruciate ligament* (ACL), thinner than the PCL, has a stronger signal intensity and presents a straight, well-defined contour of the anterior margin. *Collateral ligaments* are clearly seen on coronal and axial views with a low signal intensity. Tendons give a low signal while muscles have an intermediate signal. Separated muscle heads can be identified; intersections and fatty infiltrations are best seen on axial sections.

Vessels with rapid blood flow generally appear as dark structures in spite of their high content of mobile protons. When the signal is recorded, the excited blood protons have already moved out of the selected plane. In multiple-slice imaging, however, an intense signal may appear in some vessels due to prior excitation of adjacent layers.

Nervous structures have a relatively high signal intensity due to the fat content of myelin.

Synovial fluid has an intermediate signal on proton density and T_1 (T_1 lower than proton density) and hyperintense on T_2 and three-dimensional 40° FISP.

Fig. 2.3a–c (legend see p. 30)

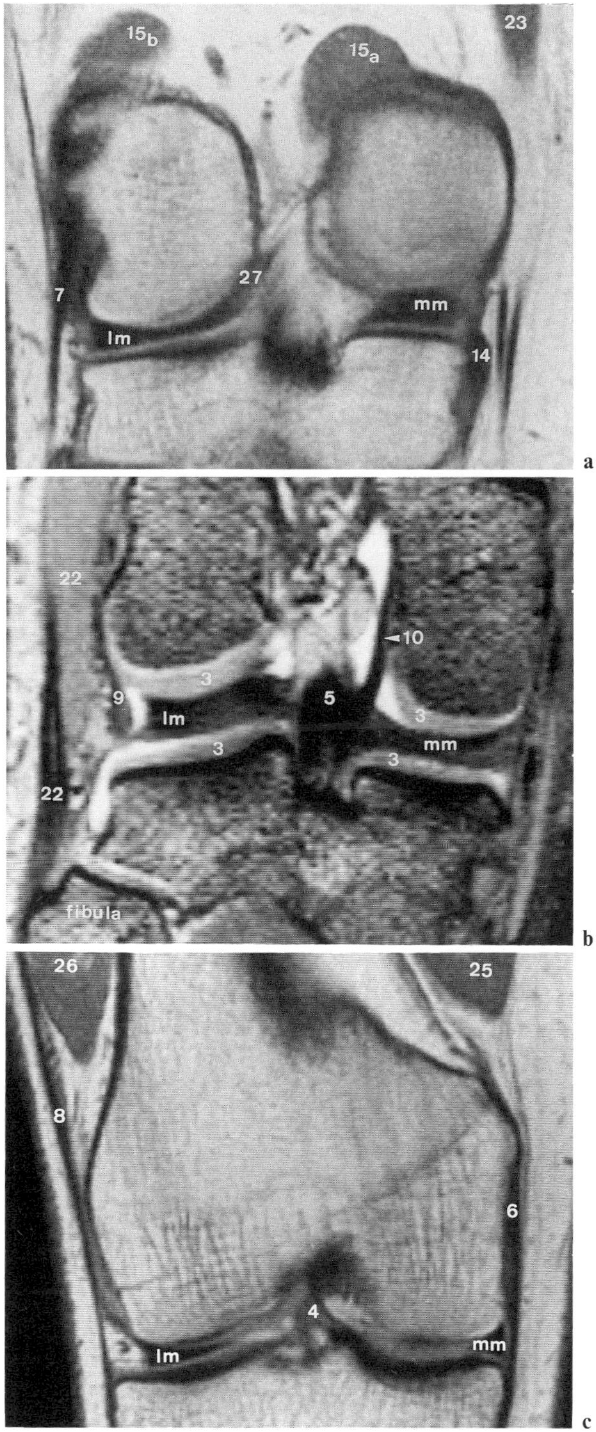

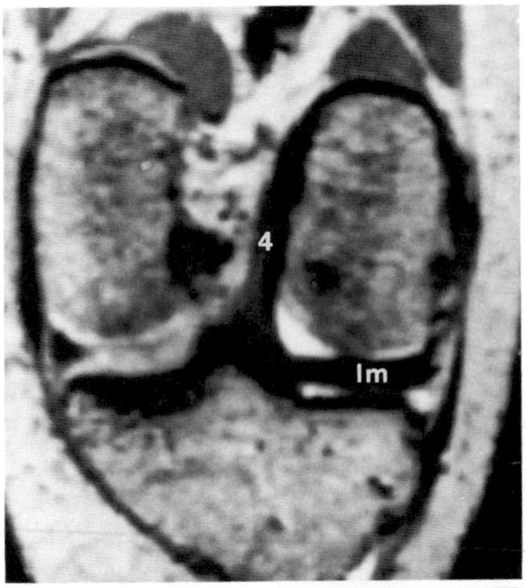

d

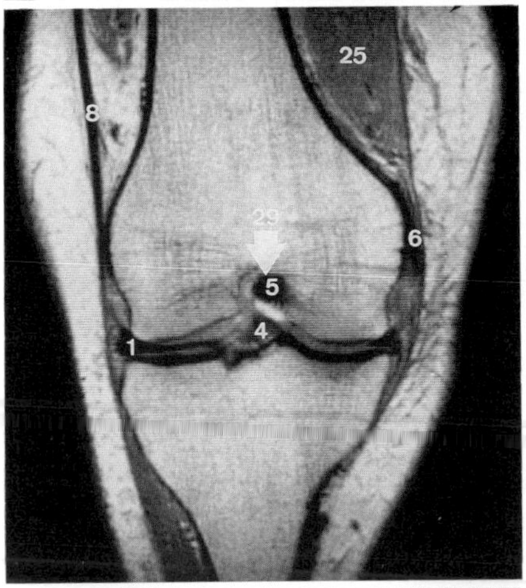

e

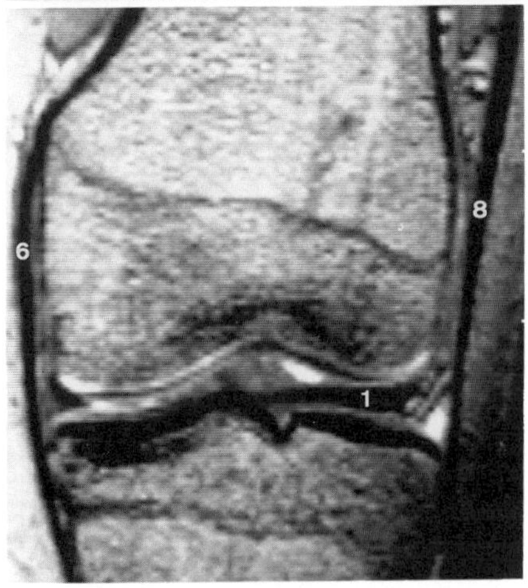

f

Fig. 2.3a–f. Coronal anatomy. **a** Section through the posterior tibiofemoral compartment (SE 2200/22). **b** Reconstruction through the posterior tibiofemoral compartment (3DFISP40 30/10). **c** Section through the ACL (SE 2200/22). **d** "Coronal oblique" reconstruction through the ACL (from a sagittal three-dimensional data set). **e** Section through the intercondylar notch (SE 2200/22). **f** Reconstruction through the anterior tibiofemoral compartment (3DFISP40 30/10). *1*, Anterior meniscus horn; *3*, hyaline articular cartilage; *4*, anterior cruciate ligament; *5*, posterior cruciate ligament; *6*, medial collateral ligament; *7*, lateral collateral ligament; *8*, iliotibial band; *9*, popliteus tendon; *10*, posterior articular capsule; *14*, semimembranosus muscle/tendon; *15a*, gastrocnemius muscle, medial head; *15b*, gastrocnemius muscle, lateral head; *22*, biceps femoris muscle/tendon; *23*, sartorius muscle; *25*, vastus medialis; *26*, vastus lateralis; *27*, Wrisberg ligament (meniscal attachment); *29*, intercondylar notch; *lm*, lateral meniscus; *mm*, medial meniscus

2.3 Sport-Induced Injuries

2.3.1 Ligamentous Disruptions

The ligament system is composed of the extraarticular ligaments, tibial and fibular collateral ligaments, and the intraarticular ACL and PCL.

Chronic instability of the knee usually signifies a lesion of the cruciate ligaments, sometimes associated with injury of the collateral ligaments. These may be secondary to progressive distension. Direct or indirect forces may result in a knee joint laxity. The position of the knee at the time of the accident plays an important role; a tackle in rugby, for example, can result in injury to PCL or ACL. A hyperextension force can rupture the PCL or, more frequently, the ACL. Posterior or anterior instability causes problems especially in such sports activities as football, hurdle race, and high jumping, which load only one leg and require strong rotations and cuttings. Two-legged activities can be better tolerated (Vanderfield and Kennedy 1985).

In acute injuries, rapid assessment of the full extent of ligament damage is extremely important for correct management of knee trauma. If not properly identified and treated, ligamentous tears predispose to further knee injury and negatively affect athletic performance (Kujala et al. 1986; Larson and Singer 1985; Levinsohn and Baker 1980; Thijn 1982; Williams 1989).

2.3.1.1 Anterior Cruciate Ligament

Rupture of the ACL, one of the most important stabilizers of the knee, is much more frequent in practicing sports (more than 95% of cases) than that of the PCL. Isolated disruptions of the ACL are the result of hyperextension, for example, a penalty kick in football, an exaggerated endorotation, or fast deceleration. Most of the time they are caused by an internal rotation of the femur on the tibia with the leg extended. They also occur very often during landing with a lightly flexed knee after a jump, as in handball, volleyball, or basketball. They may exist in an apparently innocuous injury, and their clinical diagnosis may be deceivingly reassuring; pain cannot be localized precisely by palpation. In some cases, results of the various clinical tests (e.g., anterior drawer and pivot shift tests) may be normal, and experience of the examiner is demanded, particularly with partial tears.

ACL tears are often associated with lesions of the peripheral ligamento-capsular structures and with meniscal tears. Diagnosis of lesions located in the posterior horns, particularly, may be missed by arthroscopy (Genety and Brunet-Guedj 1982; Mink et al. 1987).

MRI Appearance. The appearance of a lesion on MRI evaluation depends on the age of the lesion and on the degree of disruption. The cruciate ligaments are considered to be completely torn when their normal continuity cannot be demonstrated on thin contiguous slices (Fig. 2.4). Fluid fills the defect if the

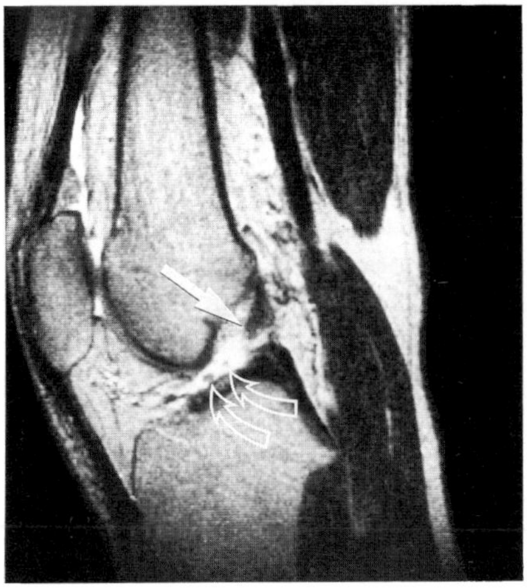

Fig. 2.4. Ligamentous disruption. Complete distal disruption of the ACL in a 25-year-old man who sustained a severe torsional valgus injury. The zone of interruption is indicated (*straight arrow*) on a 3-mm-thick sagittal T_2-weighted image (SE 2500/80); ligamentous remnants are seen (*open arrows*), surrounded by joint fluid of high signal intensity

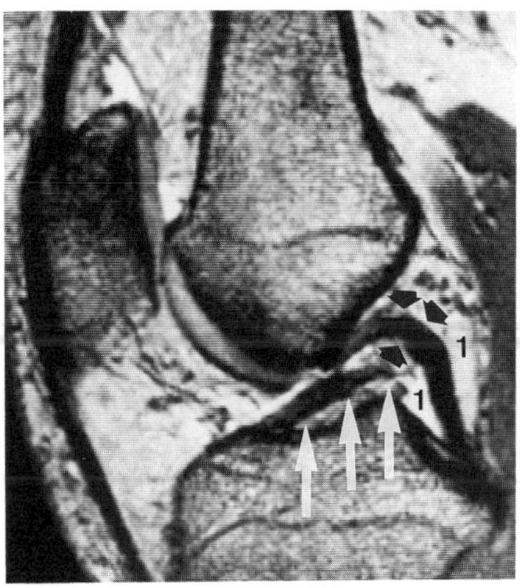

Fig. 2.5. Old ACL proximal tear in a 33-year-old man injured in a skiing accident 8 months earlier. Horizontalization of the distal fragment (*white arrows*) and buckling of the PCL (*black arrows*) are two of the indirect signs found on this isolated sagittal 1-mm-thick slice (3 DFISP 40 30/10). Note the small amount of intraarticular fluid (*1*) in the chronic phase, confirming the ancient character of the lesion

tear is not too old. Joint effusion or hemorrhage, present in nearly all cases of ligament injury, infiltrates the ligament and helps to localize the site of rupture. This precision is essential for the treatment; proximal disruptions are more common due to the weaker femoral attachment and may be directly repaired. Tears located in the middle portion of the ACL usually require tendinous or synthetic augmentation or replacement (Odensten et al. 1985). The continuity of the ligament may appear preserved with a modified architecture and signal intensity, especially in partial or intrasynovial ruptures.

The anatomical distortion of the ligament and the surrounding fat tissues is better seen on T_1-weighted images, using T_1 or T_2 sequences (on proton density images). The T_2-weighted images simultaneously obtained by T_2 sequence, demonstrate areas of higher signal intensity within the substance of the connective tissue, associated with local thickening. Three-dimensional images depict a diffusely increased signal of the full length of the ligament, corresponding to edema and fluid infiltration within the ligament sheat (Beltran 1986c; Burk 1986; Gallimore 1986; Jackson 1988; Li 1986; Mandelbaum 1986; Mink 1988; Reicher 1987; Turner 1985, 1986).

Indirect signs are also described and are helpful if three-dimensional sequence or thin T_2 spin-echo images are not available (Fig. 2.5).

- A soft tissue mass of intermediate signal intensity may be demonstrated on T_1-weighted images, usually at the proximal end of the ligament. Its signal intensity is higher on T_2-weighted images, except in the hyperacute phase.
- The anterior margin of the ACL may present an irregular and wavy contour, outlined by joint fluid of high signal intensity, on T_2-weighted images.
- Another characteristic finding is the horizontalization of the injured ligament; the remaining segment may lie free in the joint after being pulled from its femoral insertion.
- Anterior bowing of the PCL is often associated with complete ruptures, but this feature depends on patient positioning and on the degree of ligament laxity.
- Anterior subluxation of the lateral aspect of the tibia is visible, particularly in isolated tears of the ACL (Lee et al. 1988; Mink et al. 1987).

Using spin-echo sequences, T_2 sagittal images (possibly T_1, which provide less contrast with adjacent fat and fluid) have high accuracy in the detection of complete tears. An additional series of oblique (nonorthogonal) T_2 sequences may be acquired in the plane of the ACL; the combination of both sagittal and oblique orientation planes offers high accuracy in complete and partial tears, but this lengthens the examination time. Three-dimensional sequences (FISP or GRASS T_2 using a flip angle of 40°) provide the advantage of allowing an easy detection of partial ruptures by the use of very thin slices (1 mm or less). Indeed, as more than ten thin slices pass through the fibers of the ACL, a rupture of some of them can readily be seen, even if the imaging plane is not exactly parallel to the orientation of the ligament. Moreover, this fast measurement allows reconstructions along the specific plane of the ligament (even curved or angulated) and enables the differentiation of the two or three arthroscopically described bundles of the ACL (Fig. 2.6). Intraligamental bleeding and scarring are well depicted (Tyrrell et al. 1988; Spritzer et al. 1988; Shahabpour et al. 1990).

ACL grafts can also be evaluated by MRI (Fig. 2.7). Identification of both biological and synthetic graft material is possible. Complications such as inflammatory changes of the synovium, infection, rupture, or loosening of the graft can be assessed. The three-dimensional reconstructions allow visualization of the prosthesis along its precise trajectory, particularly into the femoral and tibial bony tunnels.

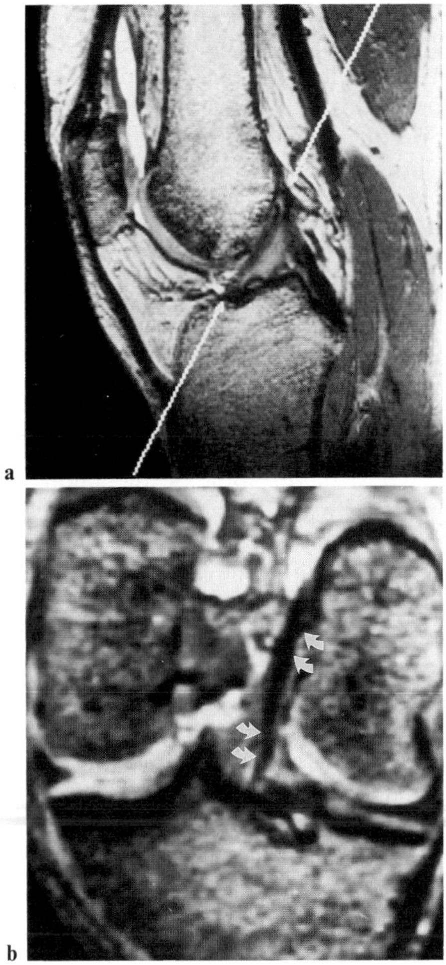

Fig. 2.6 a, b. Oblique reconstruction of ACL. **a** A 1-mm-thick sagittal view (3DFISP40 30/10) passing through the ACL, which has wavy contours and appears suspicious. **b** However, on the oblique reconstruction along the trajectory of the ACL, both femoral and tibial points of insertion are visualized, and the ligament has a normal appearance (*curved arrows*)

2.3.1.2 Posterior Cruciate Ligament

The strongest ligament in the knee is the PCL, and its rupture requires substantial energy. Injuries usually follow a posteriorly directed blow on the proximal tibia with the flexed knee and are described in motor vehicle, football, soccer, and more rarely ski accidents. The rupture may be secondary to hyperextension of the knee in some cases. The clinical tests (e.g., the posterior drawer test) are not always accurate, in case of important hemarthrosis or muscle contraction. When the ACL is intact, the tibial attachment of the PCL, most commonly injured, is difficult to visualize by the usual arthroscopic approach. A supplementary posteromedial arthroscopic portal or the use of a 70° scope is needed to diagnose a tear of the tibial insertion (Mink et al. 1987; Fowler and Messiem 1987; Zarins and Boyle 1986).

Fig. 2.7 a–c. Disruption of ACL allograft. A 35-year-old patient with persistent anterior knee instability which underwent an ACL reconstruction 10 years after a fall during a motocross race. **a** The graft is partially seen in the tibial bone tunnel (*arrowheads*) on the coronal (**a**) and sagittal (**b**) views (SE 2100/22). It shows wavy contours and areas of discontinuity on **b**, particularly in its intraarticular portion (*small arrows*). Associated lesions are also visualized on the coronal section (**a**), namely thickened medial collateral ligament (*black arrows*), torn midhorn of lateral meniscus (*curved solid arrow*), marginal osteophytes (*O*). **c** On T_2-weighted image (SE 2100/22), a fluid collection (*F*) is seen in the patellofemoral space and within the tibial tunnel. Note that the metallic artifacts due to femoral staples (*a*) are localized and not disturbing interpretation of intraarticular structures. However, small ferromagnetic remnants from surgical procedure may affect their assessment

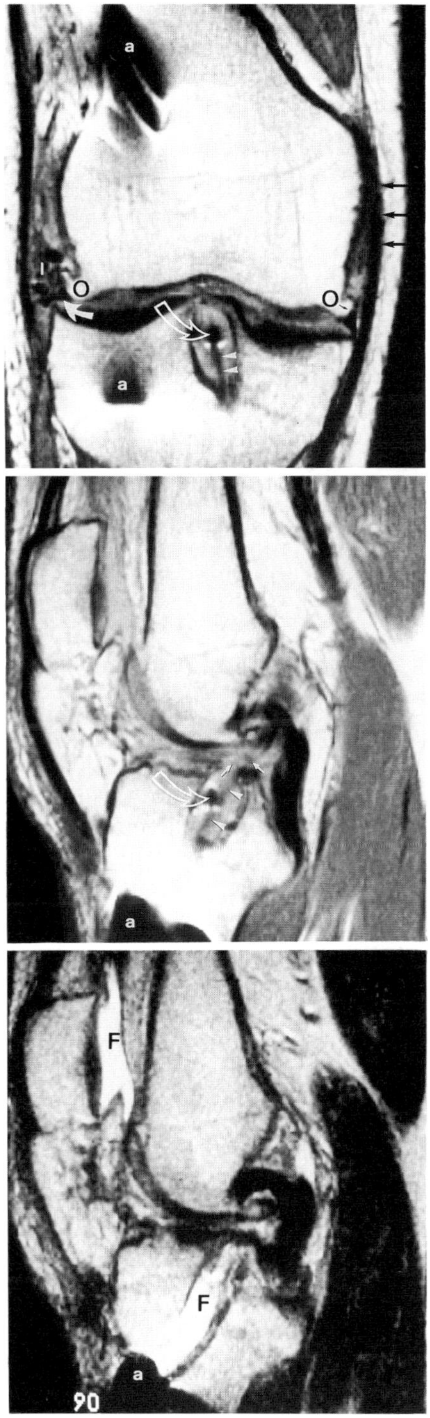

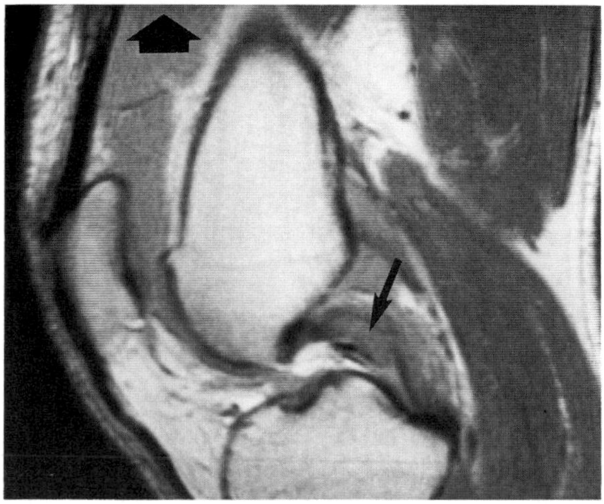

Fig. 2.8. Complete tear of the PCL in a 40-year-old parachutist with clinically suspected meniscal tear (confirmed on MR images) but negative drawer test results. The T_1-weighted sagittal view (SE 2100/22) shows the interruption of the ligament fibers with local edema of intermediate signal intensity (*thin arrow*). An important suprapatellar effusion is also depicted (*thick arrow*)

MRI Appearance. Complete and partial tears of the PCL usually appear as focal areas of increased signal intensity within the torn fibers (Figs. 2.8, 2.9). When the tear is not clearly visualized, a focal decrease of signal intensity on T_1-weighted images, brightening on T_2-weighted images, depicted in the underlying tibial bone (and corresponding to intraosseous edema or hemorrhage) may be an indirect sign of distal rupture. The diagnosis of isolated tear of the PCL seems to be essential since conservative treatment is adopted increasingly (Fowler and Messiem 1987).

2.3.1.3 Collateral Ligaments

A common disorder of the knee in athletes is the sprain of collateral ligaments, especially the medial which is typically affected in severe torsional valgus stress injuries (ski, judo, football) associated with ACL elongation or disruption.

MRI is particularly suitable for the assessment of these ligaments since they cannot be directly visualized by arthroscopy. Moreover, this operation can be a relative contraindication in the presence of massive capsular tears. Indeed, extravasation of the large amounts of fluid used during arthroscopy may result in compartmental syndromes. In severe medial collateral and ACL injuries, associated lateral meniscal tears or cartilage lesions are also described. Impaction fractures of the lateral tibial plateau and femoral condyle are not rare.

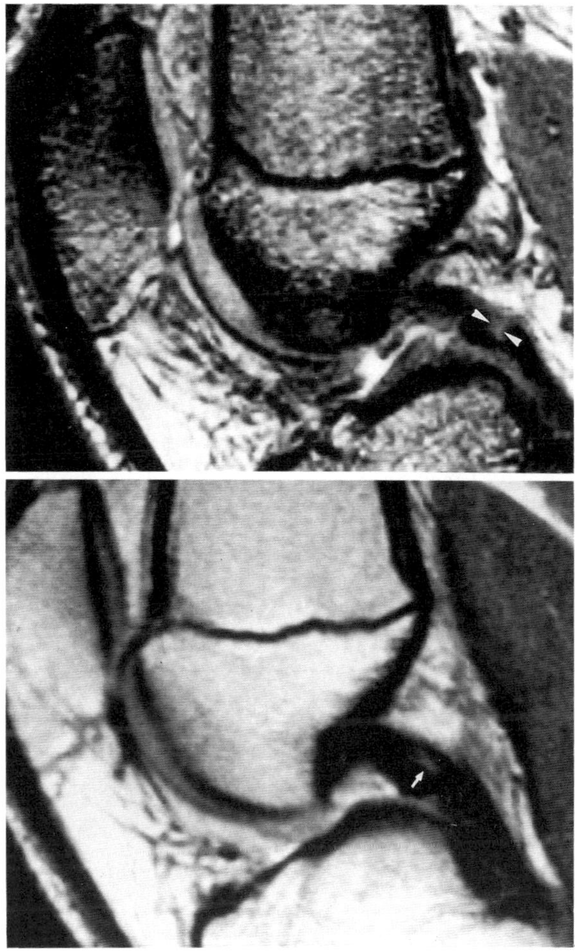

a

b

Fig. 2.9a b. Partial tear of the PCL. This 15-year-old teenager complained of posterior discomfort in the knee, which started after a football game. **a** The 1-mm-thick sagittal image (3DFISP40 30/10) shows a focal area of increased signal intensity within the PCL (*arrowheads*). Slight irregularity of the edges of the ligament is depicted. **b** On 3-mm-thick section (SE 2100/22), the zone of partial tearing is less visible (*small arrow*) within the tendon of low signal intensity, due namely to partial volume effect

MRI Appearance. Medial collateral tears are always combined with extraarticular hemorrhage and hemarthrosis; the surrounding fat tissues are infiltrated by edema and blood. Interruption of the ligamentous contours and wavy appearance are usually visualized (Fig. 2.10). Lesions of the superficial and deep portions of the ligament can be differentiated. A sprain of this ligament is detected by the presence of edema without loss of continuity (Fig. 2.11). Chronically torn collateral ligaments appear thickened and buckled (Mink 1987; Shahabpour and Osteaux 1989; Sherman et al. 1986; Turner et al. 1985).

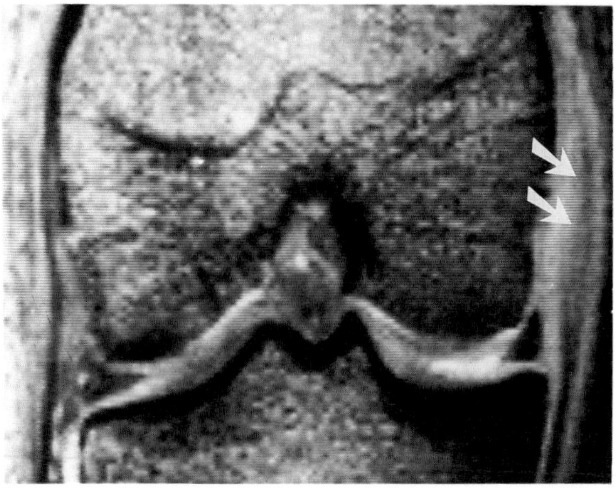

Fig. 2.10. Complete proximal disruption of the medial collateral ligament following a twisting (in valgus and external rotation) injury in skiing. Coronal reconstruction plane (3 DFISP 40 30/10) reveals a loss of continuity of the ligamentous fibers, which are proximally replaced by edema and hemorrhage (*white arrows*). This patient presented associated tears of the ACL and posterior horn of medial meniscus (not seen on this section), producing the so-called triad of O'Donoghue

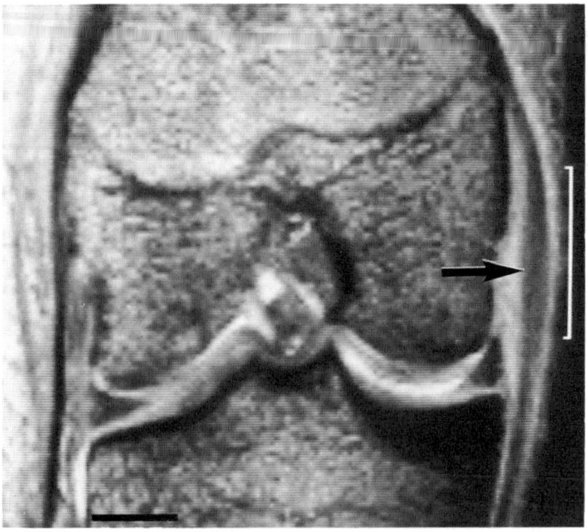

Fig. 2.11. Sprain of the medial collateral ligament in a 27-year-old soccer player presenting clinically ligamentous laxity. The normally dark and thin ligament is thickened and has a slightly increased signal intensity, corresponding to infiltration by edema (*arrow*). This ligament is medially displaced by surrounding edema, but its continuity is preserved. (Reconstruction in the coronal plane from a 3 DFISP 40 30/10)

2.3.2 Meniscal Lesions

Lesions of the menisci are one of the most frequent causes of knee pain and disability in athletes. Isolated meniscal injuries are more commonly associated with torsional forces rather than the violence of external contact. They are usually the result of a slipping, turning, or buckling of the involved joint. Sudden cuttings as in football or soccer can lead to meniscal derangment. About 50% of active soccer players get a meniscus injury; the peak frequency lies between the 6th and the 10th playing years. Meniscal tears have a high incidence in track and field and in skiing; they may also occur during triple jump or long jump in unbalanced impact (Peterson 1985; Ciullo and Jackson 1985; Cotta and Krahl 1978; Muller 1982; Ricklin 1971; DeHaven 1986).

MRI can easily detect meniscal tears, particularly those of the posterior horns, where most of them occur. These areas are acknowledged to be the most difficult for the clinician to examine. Diagnosis of these lesions by arthrography is particularly difficult for the lateral meniscus, and small lesions of the lower surface of both posterior horns, lateral and medial, may be missed by diagnostic arthroscopy.

2.3.2.1 Meniscal Tears

We can divide meniscal tears into several morphological types, for example, vertical or horizontal, longitudinal or radial. We also consider tears as acute or subacute, as well as more degenerative ones. The former are often related to one injury in young individuals, and the latter more likely found in degenerative disease. They are probably induced by repetitive microtraumata and are common in older individuals.

MRI Appearance. A meniscal tear appears as a linear or globular area of high signal intensity, which corresponds to synovial fluid infiltrating the site of rupture (Figs. 2.12–2.15). Meniscal morphology can also be modified: foreshortening of the anterior-posterior diameter, loss of sharpness of the free-edge, irregularity of the meniscal surfaces or changes of their slope (Fig. 2.16). A frequent finding in the posterior horns (particularly the medial) is an intramural (intrasubstance) tear, without extension to the articular surfaces (Fig. 2.17). On arthroscopy, these can be seen as dimples after probing. This very common feature of the adult meniscus, usually secondary to mechanical stress, may be symptomatic in some cases. Various studies on the correlation between MRI findings and those of pathology have shown that the intrameniscal area of high signal intensity corresponds to foci of mucinous degeneration. In children and young adults, however, it may correspond to normal vascular bundles (Beltran 1986a, b; Burk 1986; Crues 1987; Herman and Beltran 1988; Manco 1987; Reicher 1985, 1986, 1987; Stoller 1987; Spritzer 1988; Tyrrell 1988). Some authors have shown that accelerated preexisting meniscal degeneration predisposes to spontaneous and traumatic tears, particularly

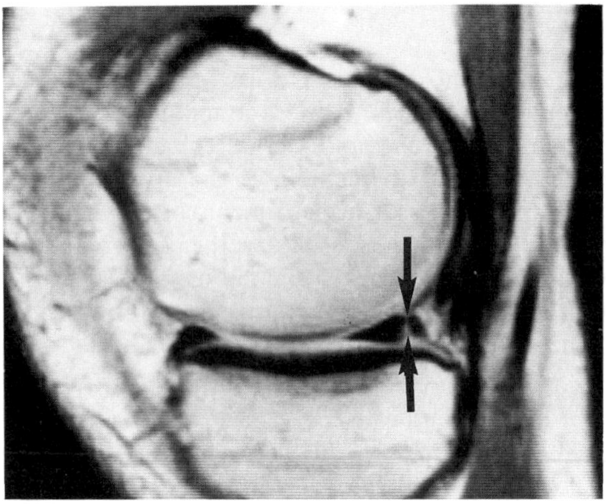

Fig. 2.12. This sagittal proton-density image (SE 2200/22) demonstrates a vertical frank tear of the posterior horn of the medial meniscus with meniscocapsular separation (*arrows*) following a twisting injury during skiing in a 20-year-old woman

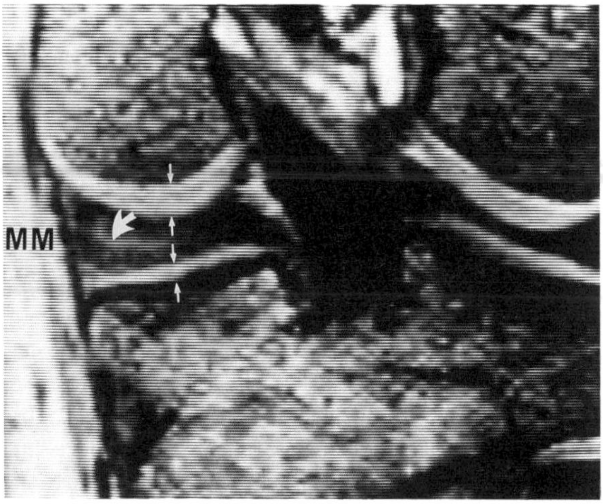

Fig. 2.13. Intrameniscal horizontal tear of the posterior horn of the medial meniscus (*MM*), well visualized after coronal reconstruction from a sagittal 3DFISP40 (30/10) data set (*curved arrow*). It does not communicate with the articular surfaces. Note the normal thickness and signal intensity of the tibial and femoral cartilage (*small arrows*)

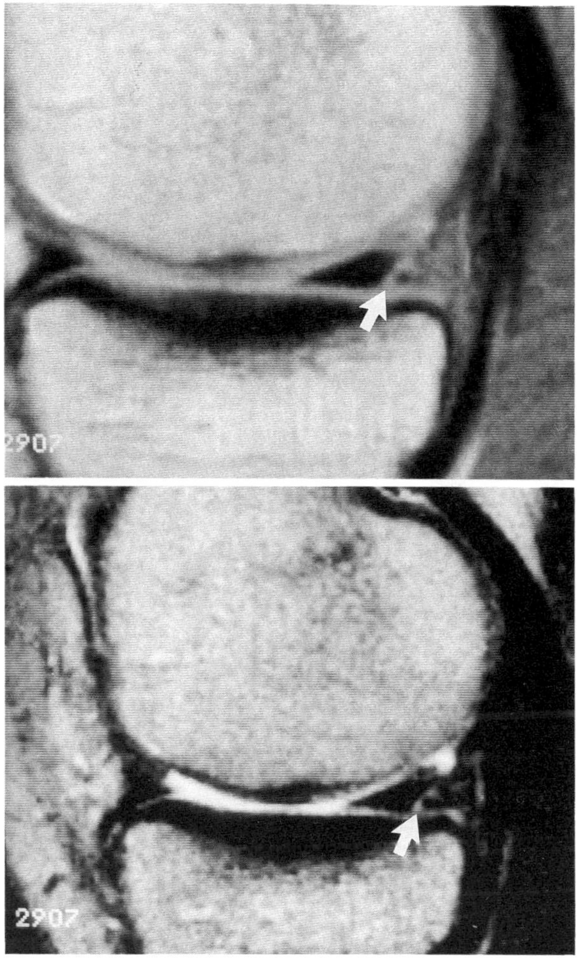

Fig. 2.14 a, b A 26-year-old man who suffered a ski injury 2 years earlier. An arthrogram was performed and considered negative. The MRI examination was demanded to exclude a medial collateral ligament tear or sprain. The sagittal T_2 sections (SE 2200/22, 70) depict an oblique tear of the inferior surface of the posterior horn of the medial meniscus (*arrow*). Due to the MR findings, the orthopedic surgeon carefully probed the posterior horn, and a small 1-cm-long beginning tear was found at its inferior margin, which was resected

those associated with recurring exercise or sports-related knee stress activities. MRI may be used to identify athletes at risk for meniscal tears (Negendank et al. 1989; Stoller 1987).

2.3.2.2 Complex Meniscal Lesions

A subtle finding on MRI is the *bucket-handle tear*. This complex lesion is vertical and longitudinal and usually occurs in the medial meniscus. The inner

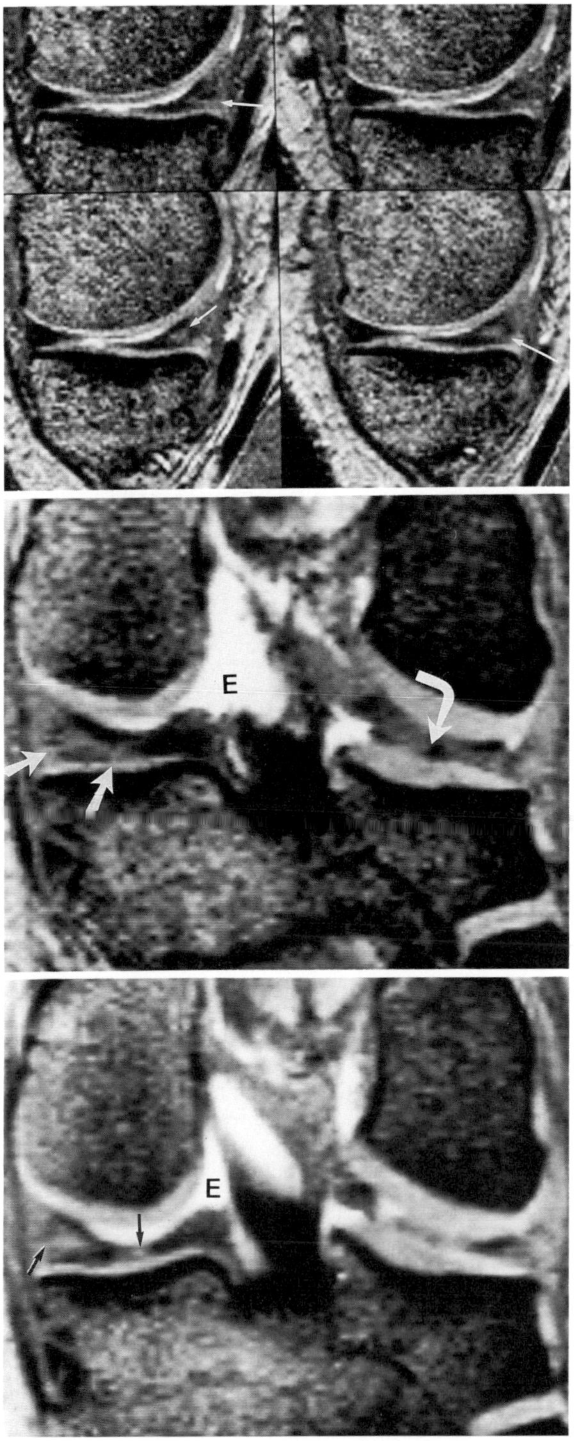

a
b
c

Fig. 2.15 a–c

fragment is displaced toward the intercondylar notch. This is classically secondary to acute injury in young athletes and should be suspected after an acute locking during effort. If the tear has extended into the anterior horn, the central fragment may be displaced anteriorly and mimic an ACL tear. On MRI examination, the sagittal planes through the intercondylar region and the coronal slices show typically the displaced fragment (Mink et al. 1987, 1988; Fig. 2.18).

The *parrot-beak tear* (Fig. 2.19), a small radial/longitudinal tear, can easily go undetected on conventional MR images. It usually occurs at the free edge, typically at the junction of the posterior and midhorns of the lateral meniscus (Mink et al. 1987, 1988; Herman and Beltran 1988). This is best visualized on axial three-dimensional reconstructions.

2.3.2.3 Discoid Menisci

The discoid meniscus is a congenital anomaly that may lead to problems; the lateral meniscus (much more often than the medial) forms a complete disc instead of having its normal C shape. There are several degrees of enlargement: mild symmetrical enlargement, assymetric hypertrophy of segments, or a massive block in the femorotibial joint space. Associations to bone and joint deformities (e.g., hypoplasia of the femoral condyle, joint widening) are also described. This disorder is typically found in childhood but may also be present in adults. Among the clinical presentations of this anomaly, one classical but not frequent feature is the snapping knee syndrome for the lateral meniscus ("click" during flexion and extension). The incidence not only of ruptures but also of intrameniscal degenerative lesions and cysts is increased, the onset occurring spontaneously or after trauma. An asymptomatic discoid meniscus should be left alone. It could sometimes be suspected on standard radiographs as a widening of the tibiofemoral joint line (Insall 1986; Williams et al. 1989).

MRI can detect and localize the tears and determine the need for surgical treatment and the choice of operative procedure.

MRI Appearance. A discoid meniscus should be suspected: (a) if the midportion between the anterior and posterior horns is seen on three or more sagittal sections, or (b) if one or two sagittal sections show triangular-shaped

◂ **Fig. 2.15 a–c.** Complex tear of the posterior horn of the medial meniscus presenting horizontal and vertical components. The topography of the tear (*arrows*) is shown on the thin (1-mm-thick) contiguous sagittal views (**a**, 3 DFISP 40 30/10) and on coronal reconstructions (**b, c**), which also demonstrate an associated joint effusion (*E*) infiltrating between meniscus and cartilage. The posterior lateral meniscus also has an abnormal shape and signal intensity on the coronal view (*curved arrow*)

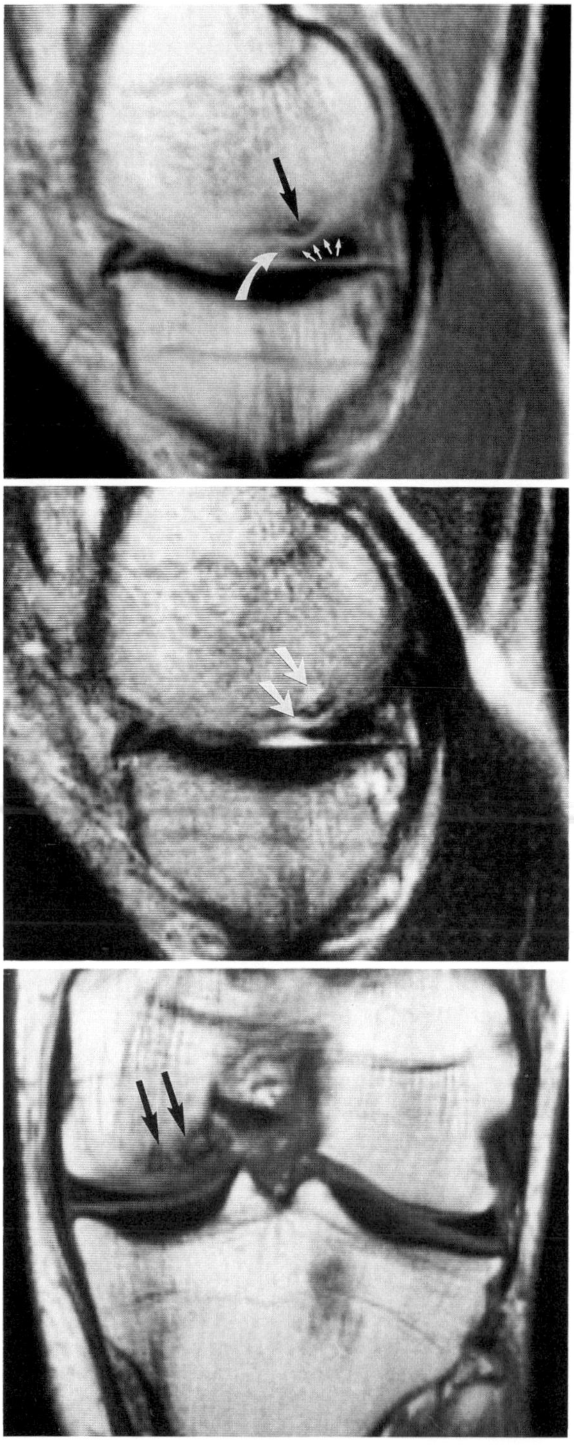
a
b
c

Fig. 2.16 a–c

◂**Fig. 2.16a–c.** A 40-year-old jogger presenting a 5-month history of diffuse pain and tenderness at the hamstrings muscles. A prior diagnosis of muscle strain was made, and the patient was treated with antiinflammatory agents and rest without relief from symptoms. Sagittal image (**a**, SE 2100/22) discloses a medial meniscal lesion with truncated free edge (*curved arrow*) and irregular superior surface (*small arrows*). An important associated cartilage lesion is diagnosed. As cartilage thickness is not accurately evaluated on SE images because of chemical shift artifact, chondral lesions can be suspected by indirect signs. A superficial subchondral area of signal loss is highly suspicious of associated chondral pathology. The low signal zone is seen on T_2 (**a, b**) and on T_1 (**c**, SE 800/15) (*black arrows*) and is surrounded by high signal on T_2-weighted image (**b**, SE 2100/70; *straight white arrows*). In this case an extensive cartilage defect was found at arthroscopy

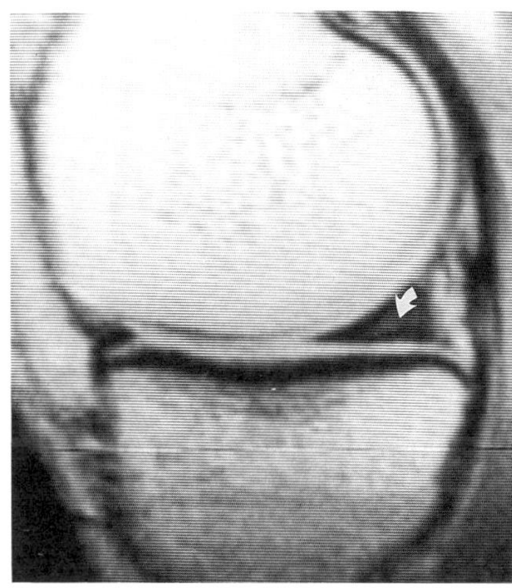

Fig. 2.17. Intrasubstance tear depicted on a sagittal section (SE 2100/22). Globular area of high signal intensity within the posterior horn of the medial meniscus (*arrow*), corresponding to an intrameniscal degenerative tear in a nonprofessional 24-year-old soccer player

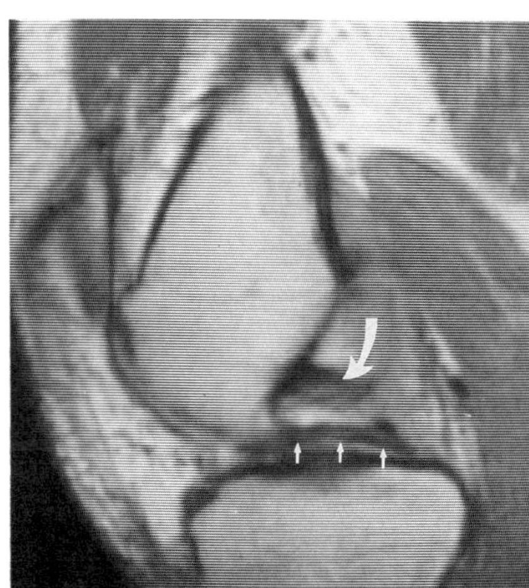

Fig. 2.18. Bucket-handle tear in a 30-year-old physical therapist who sustained an acute trauma during tennis. Clinical examination suspected an ACL tear. Sagittal proton-density image (SE 2500/22) through the intercondylar notch, displays under the proximal attachment of the PCL (*arrow*), the centrally displaced fragment of a torn medial meniscus (*small arrows*). Cruciate ligaments appeared normal at arthroscopy

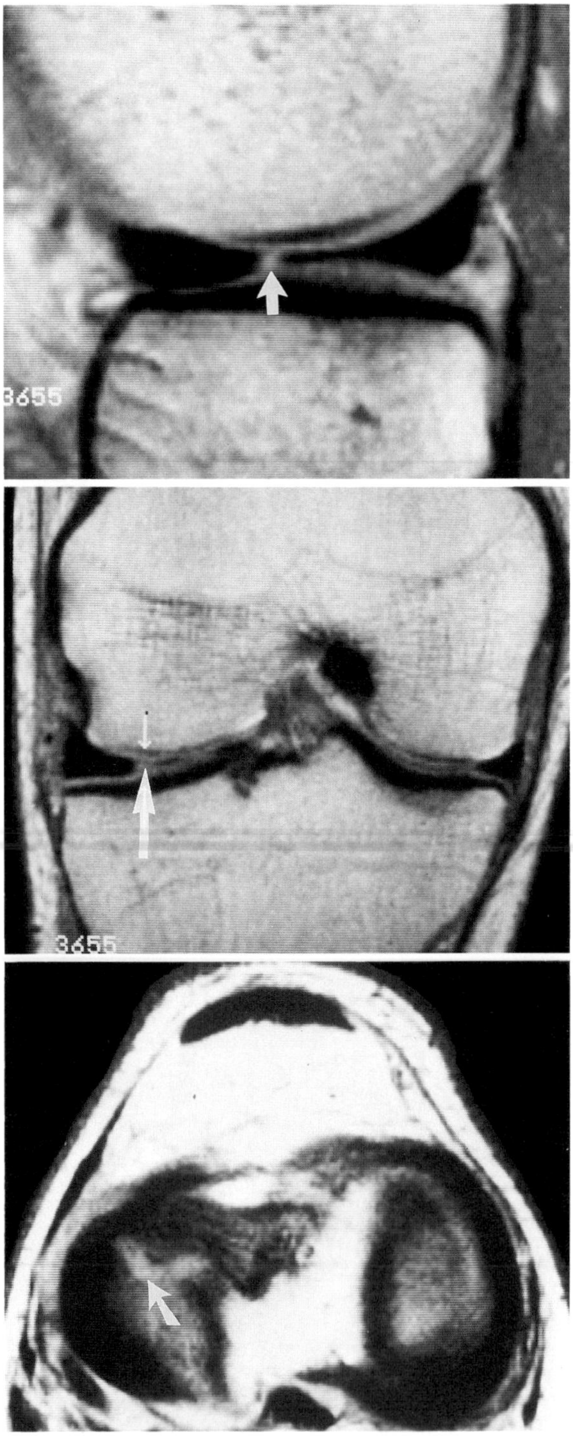

a

b

c

Fig. 2.19 a–c

◀**Fig. 2.19a–c.** A parrot-beak tear of the midhorn of the lateral meniscus in a 25-year-old soccer player after a minor trauma in valgus. **a** The vertical component is well seen on the sagittal view (SE 2100/22) as a discontinuity of high signal intensity at the junction between anterior and middle horn (*arrow*). **b** An additional coronal view (SE 2100/22) allows a better evaluation of location and extent of the tear (*arrow*). **c** A thin 3-mm-thick axial slice (SE 2100/22) depicts the horizontal component, which has an oblique course (radial tear), abuting the free edge of the meniscal body (*arrow*)

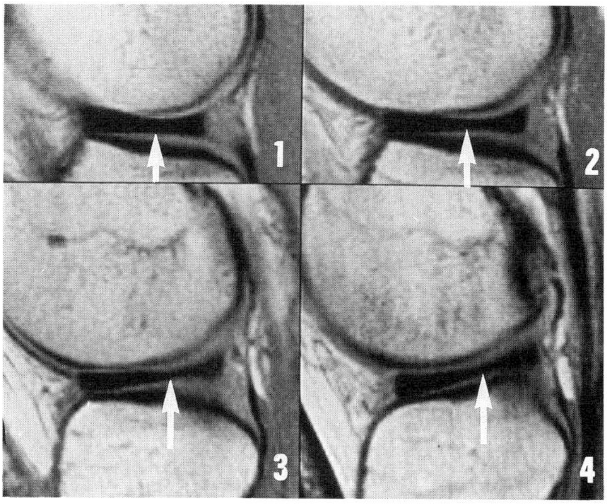

Fig. 2.20. Discoid lateral meniscus in a 29-year-old woman without history of injury, presenting a "click" during knee extension. The dysplasic meniscus was not suspected on standard X-rays and computed tomography. All four sagittal contiguous 3-mm-thick slices show a bow-tie aspect of the lateral meniscus (*arrows*). Even on the section through the intercondylar notch (*4*) there is a bridge between the normally triangular anterior and posterior horns

anterior and posterior horns (Fig. 2.20). Its full extent is best appreciated on coronal sections where the meniscus is completely interposed between femoral condyle and tibial plateau, extending into the intercondylar region. Characteristic rounded edges are often depicted (Mink 1987). The diagnosis of torn discoid meniscus with MRI (as with other diagnostic procedures) is not easy due to the numerous anatomic variations of this type of meniscus; increased signal intensity associated with enlargement are often observed (Mink et al. 1989).

2.3.2.4 Meniscal Cysts

Relatively uncommon, meniscal cysts are usually associated with degenerative tears of the menisci and are much more common in the lateral meniscus. Local

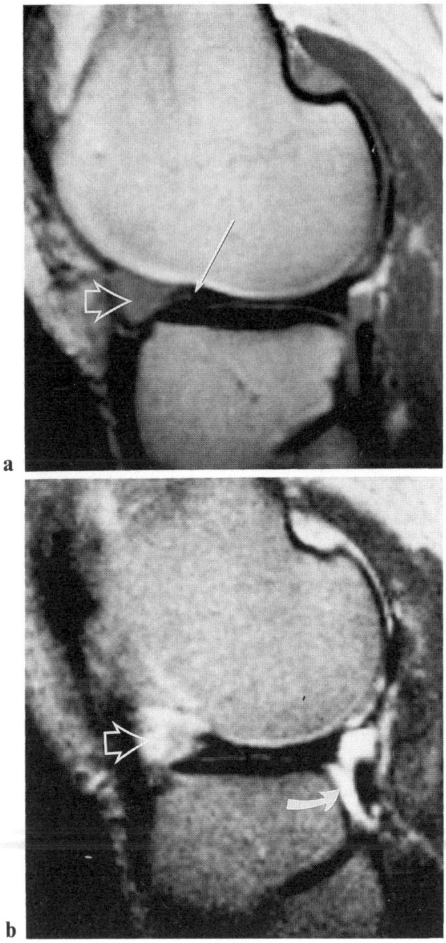

Fig. 2.21 a, b Meniscal cyst of the anterior horn (*open arrows*) typically associated with a horizontal meniscus tear (*thin arrow*). This patient with chronic anterior knee pain was clinically suspected of chondromalacia patellae. The cystic content (*open arrow*) has an intermediate signal intensity on proton-density image (**a**, SE 2500/22) brightened on T_2 (**b**, SE 2500/80). Note the slightly heterogeneous appearance of the fluid in the cyst due to its protidic content, compared to the uniformely intense signal of the distended popliteal tendon sheath (*curved arrow*)

swelling and tenderness can be detected clinically on or beneath the joint line (increasing in extension). They should be suspected in cases of spontaneous pain without a known traumatic event, appearing in full flexion under load, or commonly while turning in bed. However they may develop after traumatic episodes or be seen after meniscectomy. Other possible causes include developmental deficiency, chronic infection, and hemorrhage.

Medial meniscus cysts may present as a swelling at some distance from the joint line. They are usually asymptomatic and larger. They tend to dissect through soft-tissue planes. Meniscal cysts can rarely be suspected on conventional X-rays as a swelling of the soft tissues or by indirect signs such as like osteoarthritic changes. Arthroscopy may lead to misdiagnosis if they were not clinically suspected, and mostly when they are not distended.

Large cysts require meniscal surgery whereas smaller ones may be treated by aspiration and injection, or only be observed. It could also be interesting to

Fig. 2.22 a–c. Synovial cyst in a 20-year-old student without history of trauma, complaining of diffuse anterior knee pain which started a year earlier, after an aerobic workout. Sagittal sections through the posterior cruciate ligament (*PCL*) show a lobulated infrapatellar cystic lesion (*C*) with posterior extension. The signal intensity is intermediate on proton-density image (**a**, SE 2100/22) and high on T_2-weighted image (**b**, SE 2100/70). On an axial 1-mm-thick reconstruction (**c**, 3 DFISP 40 30/10) there is no relation with the anterior meniscal horns (*LM*, *MM*) or the patellar tendon (*P*). Arthroscopy followed by arthrotomy describe a synovial cyst developed from a fusion between medial and infrapatellar plicae

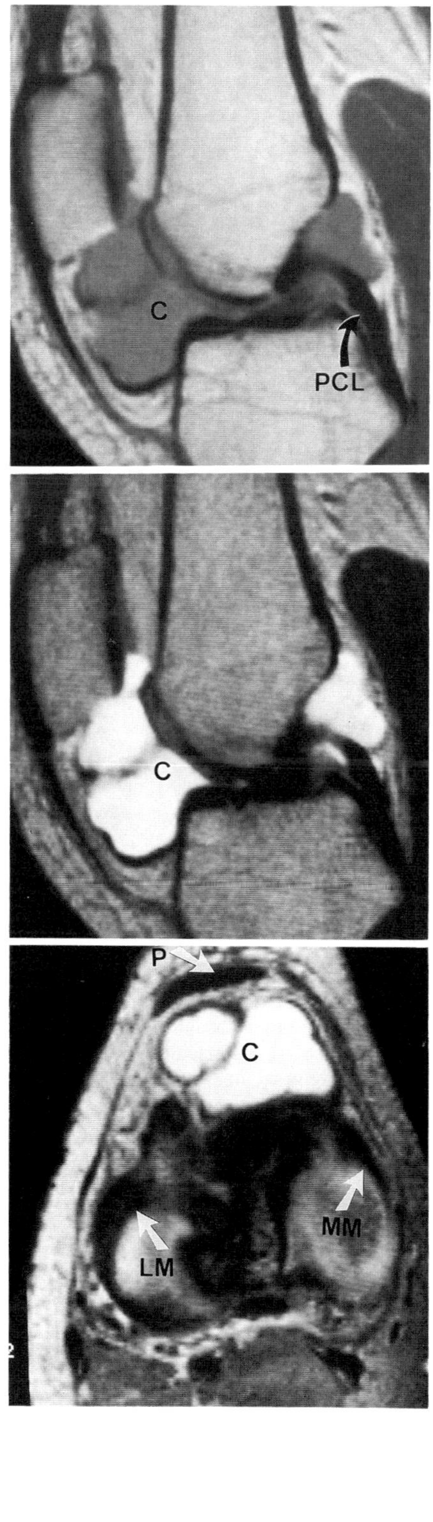

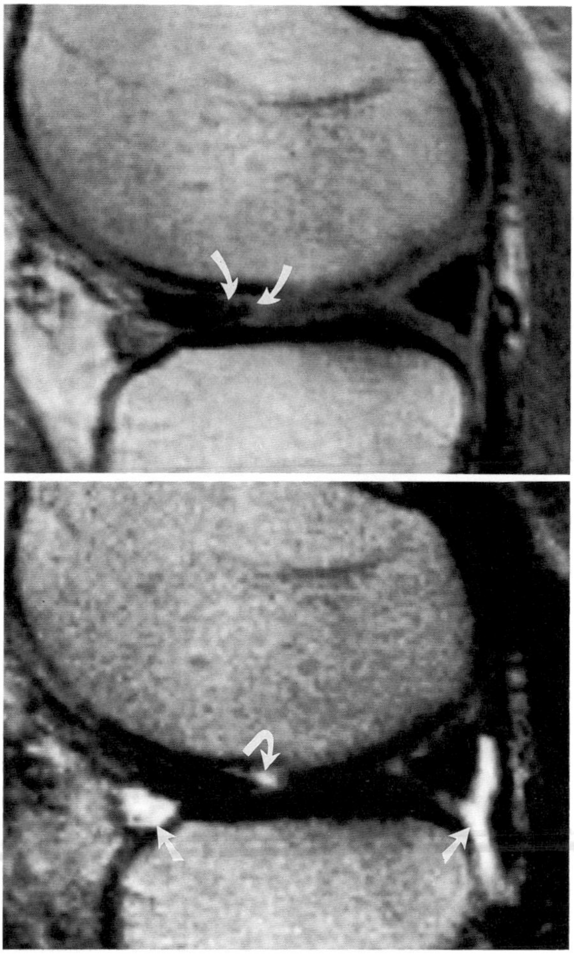

a

b

Fig. 2.23 a, b. Incomplete resection of a torn meniscus in a 41-year-old man with history of partial meniscectomy of the lateral meniscus after a trauma and presenting persistent knee pain. **a** There is a residual meniscal tear of the anterior horn with rough edges of the torn fragment (*arrows*) detected on the sagittal view (SE 2100/22). **b** A focal fluid collection is seen on the T_2-weighted image (SE 2100/70; *curved arrow*). Distension of the anterior and posterior synovial recessi is also visualized (*straight arrows*)

determine whether the meniscus is intact; plain excision of the cyst may then be enough.

MRI may incidentally discover them in examinations performed for internal derangment of the knee.

MRI Appearance. They usually present as an ovoid collection of fluid, mostly associated with a degenerative meniscal tear. They may communicate with the joint space, but in our experience the signal intensity of the cyst fluid may differ

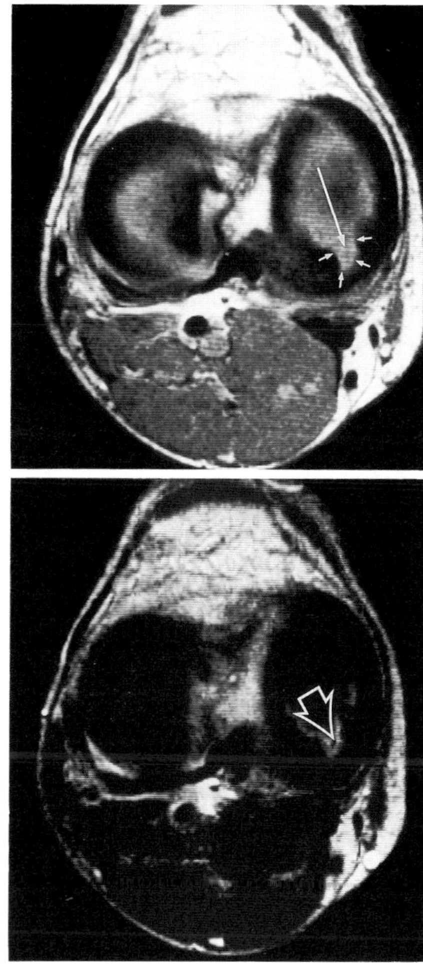

a

b

Fig. 2.24 a, b. Postoperative meniscus in a 29-year-old occasional tennis player complaining of episodes of locking and recurrent joint effusion. The patient 1 year earlier underwent an arthroscopic partial resection of the posterior horn of the medial meniscus. **a** On the axial view (SE 2100/22) there is a small body of intermediate signal (*thin arrow*) within the zone of meniscal resection (*small arrows*), consistent with an occult residual cartilaginous fragment. **b** On T_2-weighted image (SE 2100/70), the foreign body has a low signal intensity and is surrounded by a small amount of hyperintense fluid (*arrow*)

from that of joint fluid (Fig. 2.21). Sometimes they extend into the Hoffa's fat pad anteriorly. MRI is helpful to distinguish them from other disorders of the anterior fat pad and from cysts of other origin (synovial or ganglion cysts). Ganglion cysts are cystic outpouchings of the joint capsule or of other fibrous joint structures, such as ligaments, tendons, or plicae (Fig. 2.22; Burk 1986; Lee 1987; Shahabpour and Osteaux 1989).

2.3.2.5 Postoperative Evaluation

MRI is useful for evaluation of the postoperative knee. The number of patients with persistance or development of knee symptoms after complete or partial meniscectomy is significant. Among the various causes, one encounters occult

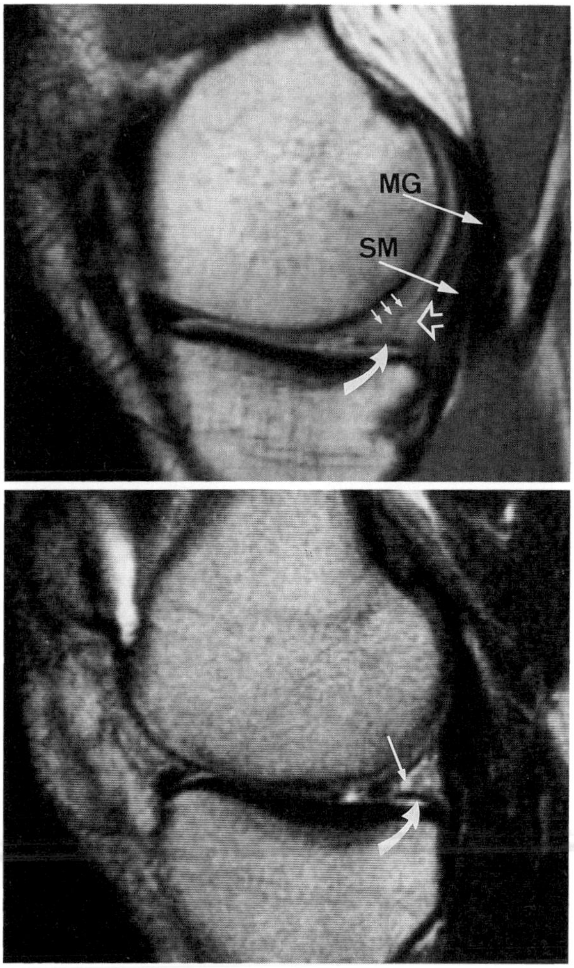

a

b

Fig. 2.25 a, b. Postoperative knee. This 29-year-old competitive volleyball player was clinically suspected of bursitis or partial tear of the pes anserinus or the tendon of the semimembranosus. He previously underwent a complete resection of the medial meniscus. Sagittal contiguous images show tissue of intermediate signal intensity on proton-density image (**a**, 2700/22; *open arrow*) with small areas of higher signal intensity on T_2 (**b**, 2700/70; *thin arrow*) in the posteromedial space. The contour of the femoral cartilage is poorly defined at this level on proton-density image (**a**, *small arrow*). The synovium bordering this space emits a low signal (*curved arrows*). On the other hand, the tendons have a normal appearance and there is no bursal effusion (*SM*, semimembranosus; *MG*, medial head of the gastrocnemius). An arthroscopy performed a few days later revealed the presence of focal cartilage lesions, associated with hypertrophic synovitis

residual meniscal fragments, incomplete resection of the torn meniscus or new tear in the meniscal remnant, and proliferation of fibrocartilaginous tissue in the area of meniscus resection. The source of postoperative pain, however, is most often chondromalacia of the femoral or tibial cartilage (softening lesions, chondral defects, or small flap tears). Fissuring of the patella, synovial scar tissue, adhesions, or chronic synovitis sometimes produce postsurgical difficulties. Common abnormalities following total meniscectomy are degenerative changes of cartilage, causing disability.

MRI Appearance. One finds shortening, signal inhomogeneity, and fragmentation of the meniscal remnants (Figs. 2.23–2.25). Regenerated menisci may resemble a meniscus in shape, but these are smaller than the previous meniscus, being only 2–4 mm in breadth. Generally, no separation with the joint capsule is seen (Mink et al. 1989). They typically present an intermediate signal intensity on three-dimensional images. The complex postoperative changes of the knee are best evaluated on three-dimensional images, particularly residual tibiofemoral or patellofemoral cartilaginous lesions (Shahabpour et al. 1989).

2.3.3 Traumatic and Chronic Pathology of the Tendons

The extensor tendons consist of the quadriceps tendon with its expansions, the lateral and medial retinaculum patellae, and the patellar tendon. Other important tendons of the knee comprise the popliteal tendon, the tendons of the hamstrings and the pes anserinus.

2.3.3.1 Acute Traumatic Injuries

Acute disruption of the extensor tendons is usually the result of indirect trauma with forceful quadriceps contraction during flexion of the knee. Complete patellar tendon ruptures are not frequent but may occur in heavily muscled young athletes or as a complication of an insertion tendinitis, particularly after repeated cortisone injections. MR images disclose a loss of continuity of the tendon with interposition of tissue of high signal intensity on T_2-weighted images (essentially hemorrhage and edema) and disruption of the normal fat planes. Quadriceps tendon ruptures usually coexist with retraction and buckling of the patellar tendon (Larson and Singer 1985; Daffner et al. 1986; Kaye and Nance 1987).

2.3.3.2 Chronic Tendinitis

Chronic lesions of the tendons include inflammatory processes often associated with partial tears, induced by repetitive microtraumata in overuse syndromes.

Most of the tendons of the knee can present exertion injuries. The chronic pathology of tendons and bursae generated by athletic activities includes:

- Patellar and less frequently quadriceps tendinitis ("jumper's knee")
- Tibial apophysitis in Osgood-Schlatter disease (in young athletes)
- Patellar retinaculitis
- Tendinitis of the hamstring tendons
- Iliotibial band syndrome
- Popliteus tenosynovitis
- Pes anserinus tendinitis and bursitis
- Prepatellar and infrapatellar bursitis
- Synovial cysts (ganglions of bone)

Diagnosis of these disorders can be based on clinical examination and ultrasonography.

MRI offers the advantages of broadening the findings of the clinical examination and providing a better soft-tissue contrast than ultrasonography. In addition, MRI permits the evaluation of an entire anatomic region, bone structures included. In addition, assessment of the precise location and size of the lesions is possible. The information given by this method can be crucial in the preoperative planning with competitive athletes (Kujala et al. 1986; Shahabpour and Osteaux 1989).

The condition termed *Jumper's knee* refers to inflammation at the proximal insertion of the patellar tendon (65% in our series) or the distal insertion of the quadriceps tendon (25%); less frequently (10%), the insertion of the patellar tendon on the tuberositas tibiae is affected. This overuse injury, resulting from intensive and repetitive stress such as in jumping, almost always concerns the jump leg. Jumper's knee is a common, painful problem among basketball players, gymnasts, triple and high jumpers, runners, and figure skaters; it also occurs in racquet sports (Bassett 1978). The symptoms are local pain on activity and well-localized tenderness. Presence of a cyst or an area of granulomatous degeneration has been described. An insertion tendinitis such as patellar tendinitis has various appearances (Figs. 2.26–2.29).

MRI Appearance. An oval or round area of high signal intensity on T_1 and T_2 images, at the location of the tendon attachment, or a focal zone of high signal intensity located only in the deep layers of the tendon insertion. These areas may correspond to edema and/or chronic partial tears due to microtraumata. Slight alterations of signal intensity within the tendinous fibers are best depicted on the thin slices obtained with three-dimensional sequences. Inflammation of the entire patellar tendon is easily diagnosed by MRI. A complete tendinitis appears as a thickening of the tendon with a central area of increased signal intensity affecting the whole length on T_1, T_2, and three-dimensional images (Fig. 2.30; Roels 1978; Turner and Prodromos 1986; Beltran 1987; Reicher 1987; Mink 1987; Bodne 1988).

Other forms of tendinitis, tenosynovitis, and bursitis (Kujala et al. 1986) can also be examined by MRI. While *tendinitis of the hamstrings* is frequently

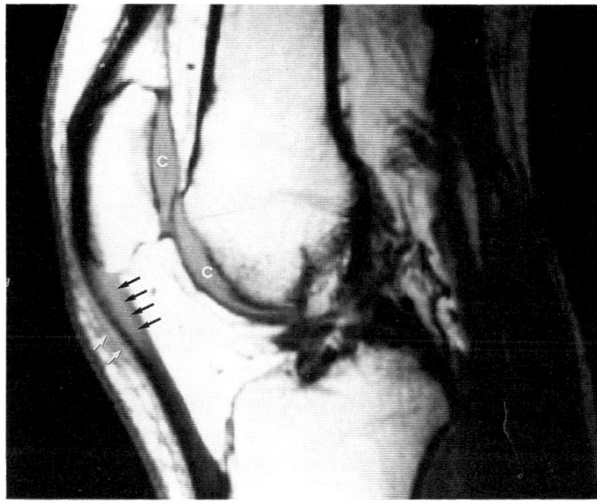

Fig. 2.26. Insertion tendinitis of the patellar tendon encountered in a 25-year-old competitive volleyball player with chronic pain under the patella, increasing after exercise. On the sagittal proton-density image (SE 2500/22) the tendinous insertion into the patella is replaced by an extensive zone of high signal intensity (*black arrows*). Focal infiltration of the subcutaneous tissue is noticed (*white arrows*). The patellofemoral cartilage (*c*) looks perfectly normal at this level

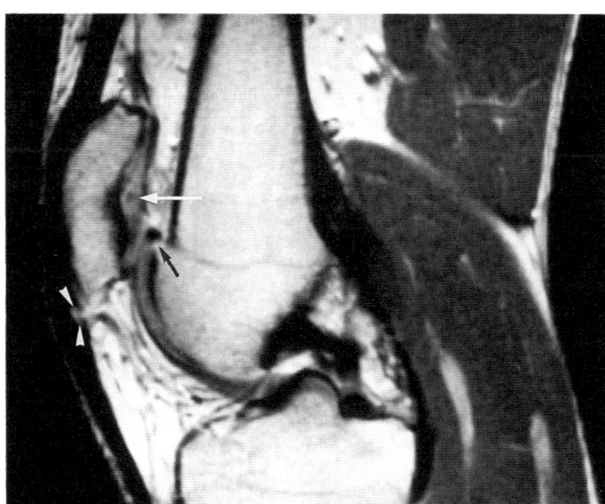

Fig. 2.27. Jumper's knee with partial tear of the patellar tendon and associated chondromalacia patellae (*white arrow*). The tearing appears as focal areas of discontinuities in the proximal portion of the tendon with higher signal intensity (*arrowheads*). Note the prominent shape of the superior margin of the femoral condyle (*black arrow*), often associated with chondromalacia. (Drs. J. Malghem and B. Maldague; Cliniques Universitaires St Luc, Université Catholique de Louvain, Brussels)

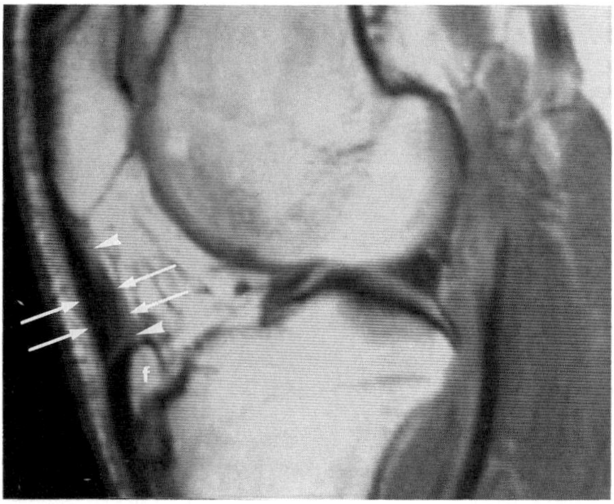

Fig. 2.28. Chronic patellar tendinitis in a 22-year-old long-jumper with history of Osgood-Schlatter disease presenting infrapatellar pain and tenderness. There is an avulsed osseous fragment (*f*) at the tibial tuberosity on this proton-density image (SE 2200/22). A slight thickening of the patellar tendon is observed (*arrows*), associated with areas of higher signal intensity in the deep layers (*arrowheads*). The infrapatellar space is filled by a small fluid collection of intermediate signal intensity

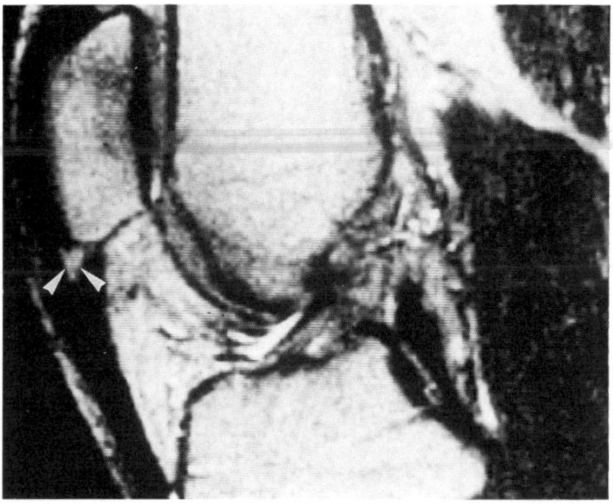

Fig. 2.29. Focal degenerative tendinitis in a 26-year-old basketball player. A central zone of intense signal (*arrowheads*) within a markedly enlarged tendon at the patellar attachment (SE 2200/80) proved to be mucoid degeneration at surgery

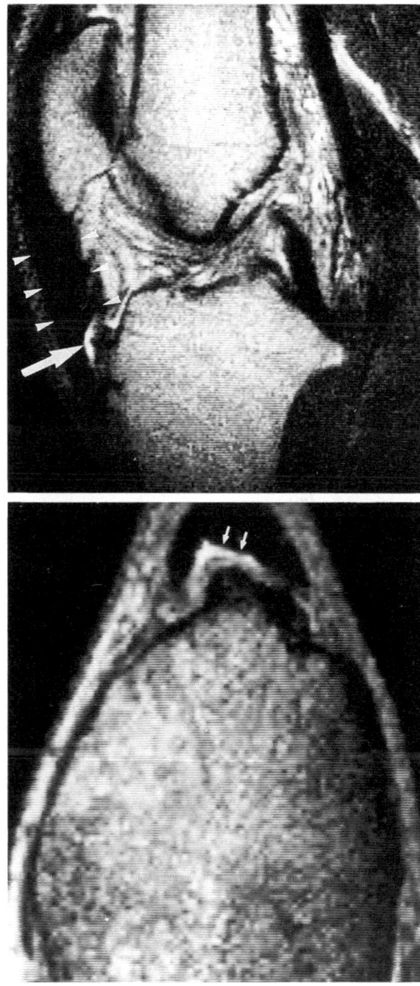

a

b

Fig. 2.30 a, b. Extensive tendinitis of the whole patellar tendon coupled with chronic tear of the attachment into the anterior tibial tuberosity, ascertained in a top athlete. He presented knee pain in childhood attributed to Osgood-Schlatter disease. **a** A thin sagittal T_2 section (SE 2200/80, 3 mm) shows diffuse thickening of the entire tendon (*arrowheads*) with a small focus of increased signal near the tibial insertion (*arrow*), probably corresponding to chronic tearing with local inflammation. **b** Axial reconstruction from a 3DFISP40 sagittal data set (1 mm) allows easy evaluation of the extent of lesions and the degree of thickening (*small arrows*)

seen in sport activities such as lacross, high jumping, and triple jumping, the *iliotibial band syndrome* occurs more often in runners, hammer throwers, and those in racquet sports. In sprint, the shortened length and increased barking of indoor surfaces is often held responsible for the iliotibial band syndrome. In distance running, hills must be avoided since these injuries may be accentuated by repetitive uphill running (Michael and Mattheus 1985; Orava 1978; Ciullo and Jackson 1985). In contrast with the iliotibial band syndrome, *popliteus tenosynovitis* is frequently a result of "jamming on the brakes" in downhill running (Mayfield 1977). *Pes anserinus bursitis*, on the other hand, seldom occurs in runners but is commonly seen in the breaststroker swimmer. With MRI, fluid in the tendon sheath of the semimembranosus or the medial head of the gastrocnemius can be differentiated from popliteal cyst (commonly called

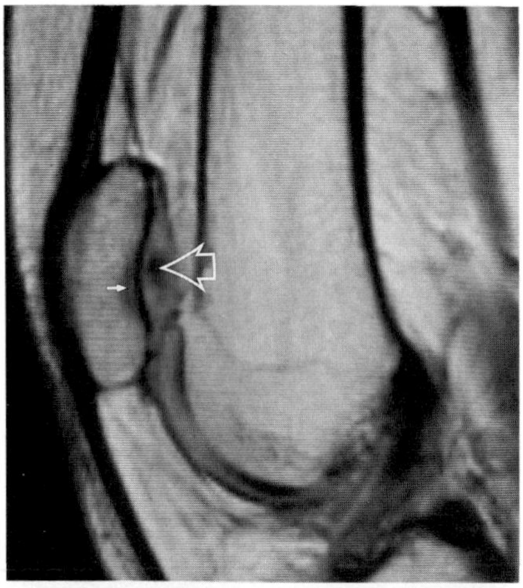

a

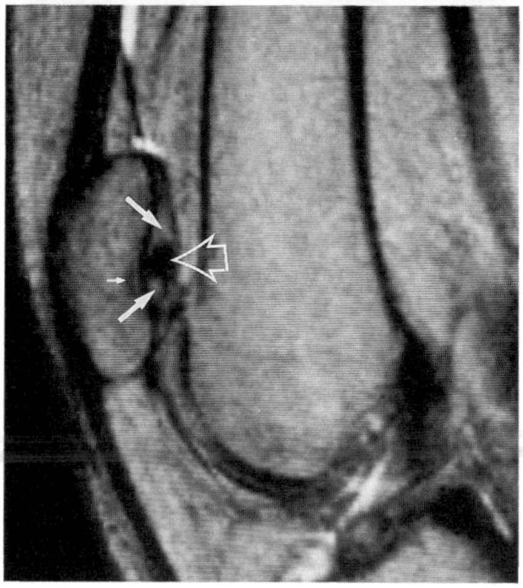

b

Fig. 2.31 a, b. Softening lesion of patellar cartilage in a 22-year-old female jogger presenting severe pain under the knee cap, particularly after exercise. Sagittal views (SE 2200/22, 80) show a focal round area of low signal intensity in the middle of the swollen cartilage (*open arrow*). Signal intensity of the surrounding cartilage slightly brightens on T_2-weighted image (*large arrows*). A small area of subchondral sclerosis (*small arrow*) is also visible

Baker's cyst). An avulsion injury can occasionally be detected at the posteromedial corner of the tibial plateau at the distal point of attachment of the semimembranosus (King 1986; Yao and Lee 1989). *Infrapatellar bursitis* is also seen in wrestling, a sport which requires almost constant intimate contact with an opponent and the mat (Snook 1985). Tenosynovitis and bursitis are characterized on MRI by collections of fluid of high signal intensity on T_2-weighted images, filling the tendon sheaths and joint recessi, respectively.

2.3.4 Disorders of the Cartilage

Patellofemoral pain syndrome usually occurs in young athletes (runners, gymnasts, tennis and squash players) as a consequence of overuse.

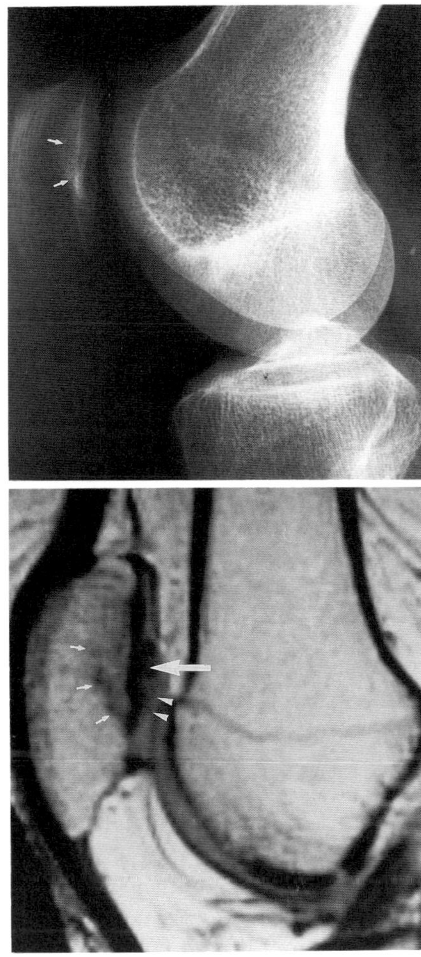

Fig. 2.32 a, b. Posttraumatic chondromalacia patellae in a 29-year-old female cyclist with anterior knee pain since a fall. **a** Lateral X-ray view shows subchondral sclerosis at the lateral patellar facet (*arrows*). **b** MRI examination ruled out any meniscal lesion or pathological plica and allowed detection of a small lesion of the patellar cartilage. Arthroscopy was delayed, and the patient showed release of symptoms after a few months of conservative therapy. On the sagittal section (SE 2100/22) passing through the patellar crest, a focal intracartilaginous low-signal area is seen (*large arrow*), associated with irregular edge of the patellar cartilage (*arrowheads*) and slight subchondral sclerosis (*small arrows*)

2.3.4.1 Chondral Lesions

Chondromalacia patellae, literally a softening of the patellar cartilage, is the diagnosis commonly applied to the syndrome of chondral lesions. It is considered as a frequent cause of internal derangment of the knee; occasionally, the cartilage of the femoral condyles and the tibial plateaus are affected (Ficat 1987; Mink 1987; Resnick 1989; Shahabpour 1989). The usual complains are dull and disturbing pain around or behind the patella, increasing in flexed knee activities. Besides patellar instability, the releasing factors are direct trauma (for example, fall on the knee) and repeated microtrauma. Except for cases of patellar instability or morphological abnormalities, long-term results of surgical treatment are often disappointing, and conservative therapy is generally recommended. Clinical examination is not always specific,

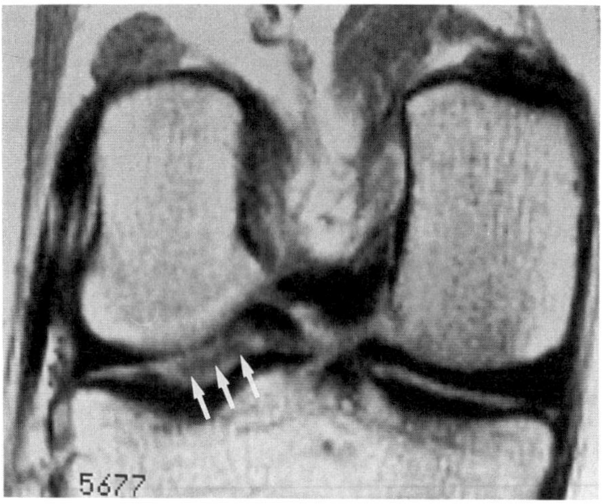

Fig. 2.33. Tibiofemoral cartilage changes can sometimes be encountered on spin-echo images (SE 2100/22). On this coronal section, a linear zone of low signal intensity is seen within the medial cartilaginous portion of the lateral tibial plateau (*arrows*)

and there is no correlation with arthroscopic staging. Conventional radiographs provide only indirect signs of severe lesions, i.e., small irregularities or sclerosis of the posterior surface of the patella (Fig. 2.32a). Arthrography combined with computed tomography can detect lesions in the late stages. Until now, arthroscopy has been considered as the diagnostic method of choice (Maldague and Malghem 1978; Thijn 1982; Boven et al. 1982; Malghem and Maldague 1989).

MRI allows a noninvasive and accurate evaluation of early pathological changes. Aspect, topography, and especially the thickness of the cartilage, which is more difficult to measure with arthroscopy, can be assessed. (Gylis-Morin 1987; Konig 1987; Wojtys 1987; Yulish 1987; Reiser 1988; Tyrrell 1988; Shahabpour and Osteaux 1989). In addition, other causes of femoropatellar pain can be excluded, e.g., anterior meniscal lesions, pathological plicae, or more rarely Hoffa's disease or patellar tendonitis. The most commonly affected area is near the median ridge, where the patella is thickest. The patellar lesions are occasionally combined with alterations in the patellar facies of the anterior femoral condyles.

MRI Appearance. MRI permits the staging of chondral lesions. In stage 1, arthroscopically detected, softening lesions usually appear as focal hypointense areas within the cartilage on T_2 images and on three-dimensional images; the articular surface appears intact but often shows a slight elevation, probably secondary to edema. The tibial and femoral locations are better evaluated on three-dimensional images (due to the diminution of the chemical shift artifact and partial volume effect) (Figs. 2.31–2.33). In stage 2, hypointense foci

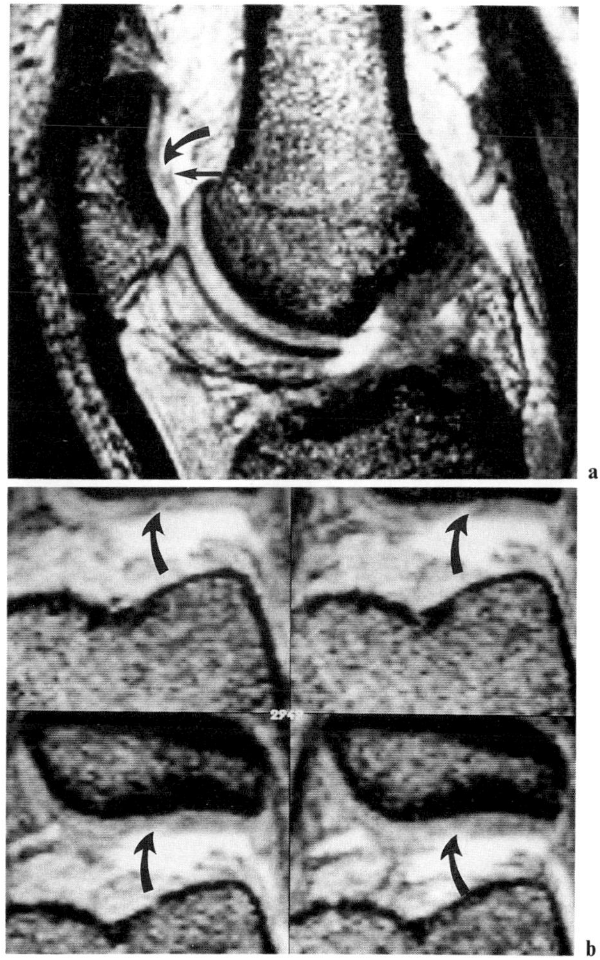

Fig. 2.34 a, b. This traumatic chondral flap tear of the lateral patellar facet (*straight arrow*) was disclosed thanks to the thin sagittal sections (**b**, 3 DFISP40 30/10). The associated extensive cartilage defect (*curved arrows*) is well seen on the contiguous axial reconstructions (**b**). The patient had a history of a fall 1 year earlier and presented knee pain after long-distance running

surrounded by high signal on T_2-weighted images, generally correspond to small fissures, sometimes associated with flap or crab meat lesions at arthroscopy. The roughened surface with fissures or flaps is best recognized on the thin three-dimensional slices (Fig. 2.3.4). In stage 3, superficial and deep defects with exposure of subchondral bone are easily demonstrated on MR images, better with three-dimensional sequences regarding femorotibial location (Figs. 2.35–2.37; Yulish et al. 1987; Mink and Deutsch 1989; Lehner et al. 1989; Shahabpour and Osteaux 1989; Shahabpour et al. 1989).

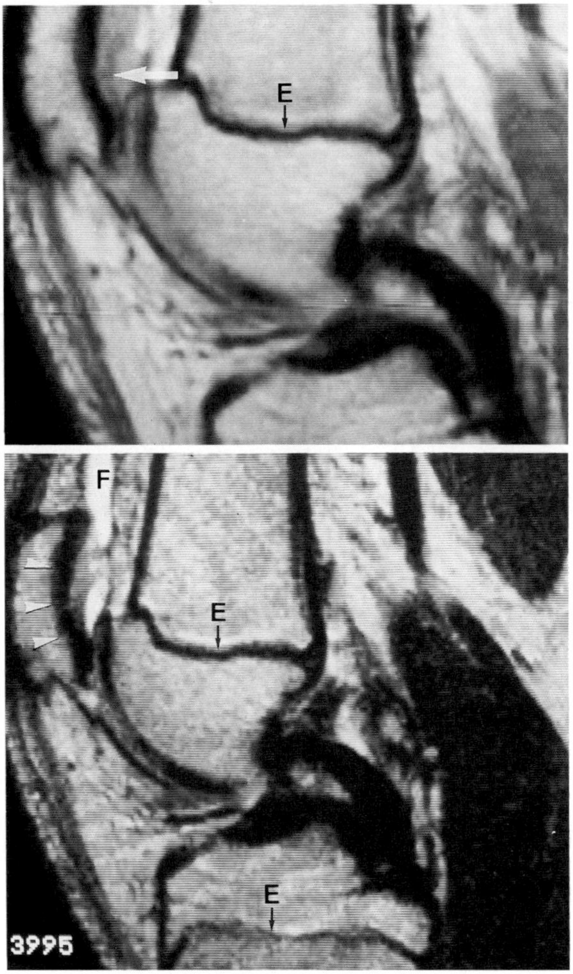

a

b

Fig. 2.35 a, b. Traumatic cartilage destruction following patellofemoral dislocation. A 13-year-old hockey player presented a history of important fluid effusion, without known injury. An episode of patellar dislocation could explain the clinical findings. On sagittal T_2 images (SE 2100/22, 70) an extensive cartilage defect of the medial patellar facet (*arrow*) with thickening and deformity of subchondral cortical bone (*arrowheads*) is depicted. Note the fluid effusion (*F*) in the suprapatellar pouch and the epiphyseal lines (*E*) well seen in this young girl

Fig. 2.36 a–c. Patellofemoral degenerative changes. This 55-year-old patient, who underwent ▶ a complete rupture of posterior cruciate ligament while skiing, presented with patellofemoral complaints. **a** Cartilage of both articular surfaces and underlying cortical bone (*white arrows*) are thinned and have irregular outline and heterogeneous signal intensity on the sagittal proton-density image (SE 2100/22). **b** On a sagittal 1-mm-thick section (3 DFISP 40 30/10), tiny irregularities of the cartilage contours are visible (*small black arrows*). **c** Axial reconstruction confirms the presence of small fissures and allows their precise location (*black arrows*). Note a degenerative lesion of the patellar tendon, presenting a focal increase of signal intensity on the sagittal views, better seen on the thin three-dimensional section

Fig. 2.36 a–c

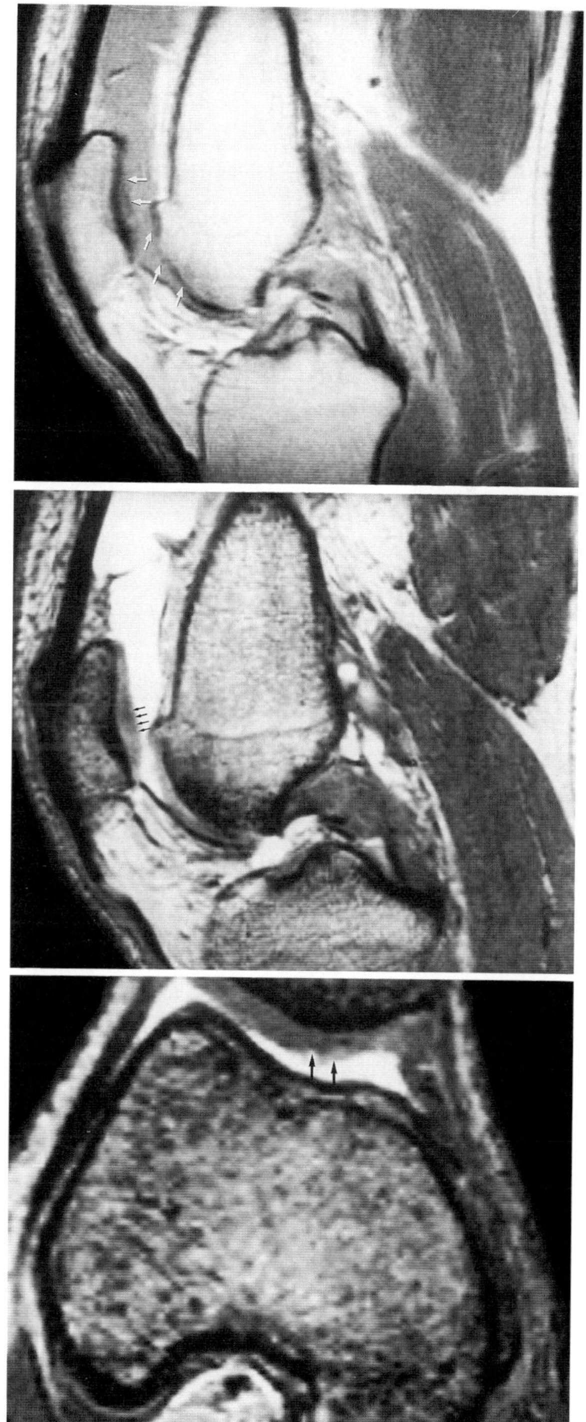

a

b

c

a

b

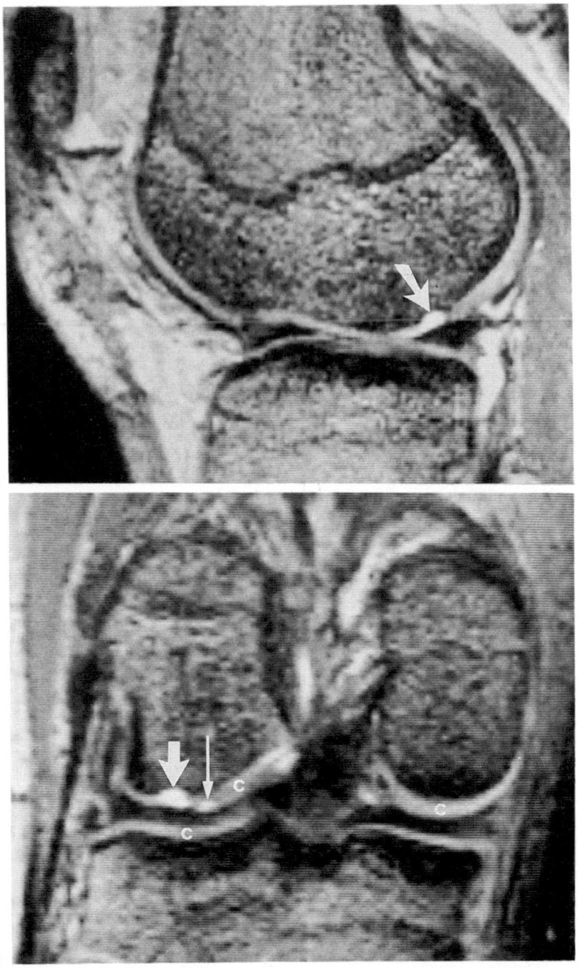

Fig. 2.37. **a** Sagittal 1-mm-thick section (3 DFISP 40 30/10) showing a deep cartilaginous defect of the lateral femoral condyle (*large arrow*) with exposure of subchondral bone in a 28-year-old patient. **b** The coronal reconstruction shows an additional, more superficial cartilage defect (*thin arrow*). The surrounding cartilage (*c*) has a normal thickness and signal intensity. The zones of defect are filled by fluid of high signal intensity

2.3.4.2 Traumatic Patellofemoral Dislocation

Dislocation or subluxation of the patellofemoral joint can occur after a major injury in a normal knee or after a lesser trauma if there is a predisposition, such as malalignment. It is usually easy to diagnose by the clinical examination. However, this injury is frequently associated with an osteochondral or chrondral fracture, resulting in development of loose bodies, which can induce recurrent knee complains (Figs. 2.35, 2.38). Partial or complete tear of the

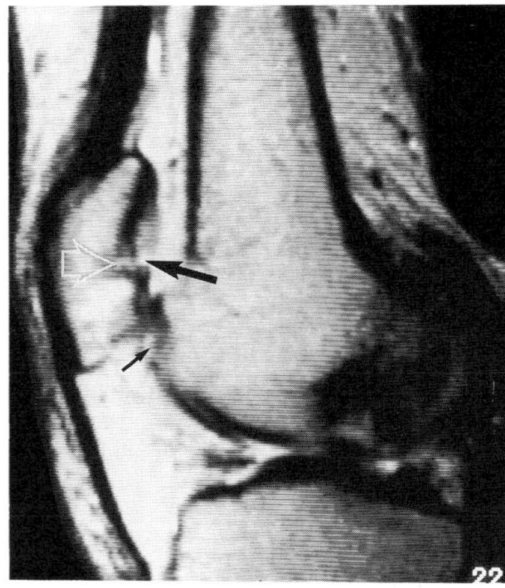

Fig. 2.38. Osteochondral fracture in a nonprofessional gymnast. An evident cortical break (*open arrow*) is associated with a patellar chondral defect, filled by high signal intensity fluid, on this T_2-weighted image (SE 2200/80; *large arrow*). The facing anterior femoral cartilage is also injured (*small arrow*)

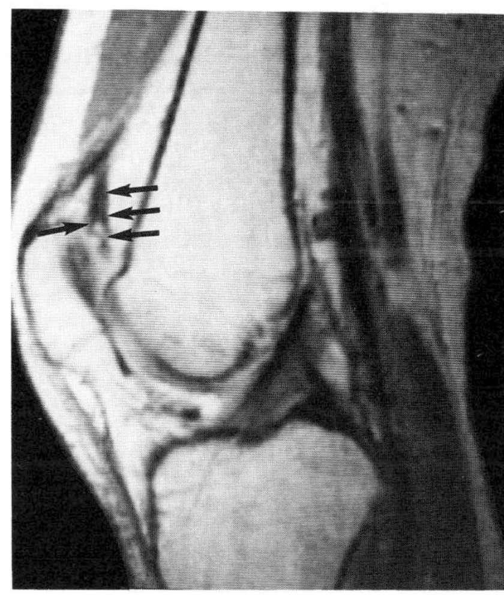

Fig. 2.39. Pathological plica. This 35-year-old occasional tennis player presented with chronic knee pain for 18 months, with recurrent reflex sympathetic dystrophy. The clinical diagnosis was chondromalacia patellae, but there was no pain relief after physiotherapy. On this sagittal section (SE 2200/22) through the patellofemoral joint, a thickened mediopatellar plica with irregular edges was discovered (*arrows*). Relief from symptoms was observed after surgical resection of the plica

retinaculum patellae are depicted by MRI, and the presence of intraarticular loose body can be disclosed.

2.3.4.3 Other Causes of Anterior Knee Pain Syndrome

The medial patellar plica is a synovial fold, beginning on the medial articular synovium and inserting into the synovium covering the infrapatellar fat pad. The medial plica is a rare but important cause of anteromedial knee pain. It becomes symptomatic when thickened and replaced by fibrosis, namely after direct or repeated trauma. An inflamed plica may impinge on the medial femoral condyle or the medial facet of the patella during flexion/extension and produce chondromalacia. Symptoms may include catching, clicking, and giving way.

MRI can precisely localize the origin of the pain by detection of pathological signs of the inflamed plica and analysis of associated cartilage lesions. The morphological characteristics are better evaluated than with computed tomography and generally in more physiological conditions, without artifical articular distension.

MRI Appearance. The fibrous band is easily depicted when surrounded by intraarticular fluid; in some cases, assymetry in the medial wall of the patellar pouch is a helpful sign (Fig. 2.39).

2.3.5 Bone Injuries

Not only soft tissues of the knee can be damaged by sport. Bone injuries also occur frequently, sometimes associated with cartilaginous lesions. The main conditions are: osteochondrosis, osteochondritis dissecans, and Osgood-Schlatter disease; osteochondral injuries; occult fractures; tibial and femoral fractures of the knee; and posttraumatic osteonecrosis. While conventional X-ray images allow a good evaluation of mineralized bone elements, MRI can directly image the bone marrow and is the most sensitive imaging technique to assess pathological processes affecting this area.

2.3.5.1 Osteochondritis dissecans

Those presenting with osteochondritis dissecans, generally children and young adults, show pain, effusion, and intermittent locking of the knee. This process affects a fragment of articular cartilage with underlying subchondral bone, which is partially or completely separated from the subjacent bone. The classical site is the lateral aspect of the medial femoral condyle; other locations include the medial aspect of both condyles and rarely the patella. Symptoms such as giving way, catching, and locking suggest a separation of the fragment. The etiology remains controversal. This lesion can be related to acute or repeated minor trauma. A common traumatic process is an impingement of the anterior tibial spine against the lateral side of the medial femoral condyle while the knee is in full extension. Local vascular bone infarction is another etiologic theory.

It is important to evaluate the nature of the articular cartilage overlying the involved bone area. If the cartilage is intact (in situ loosening), healing of the lesion may occur with time and decreased stress. In case of interruption in the continuity of the cartilage, synovial fluid may infiltrate through the break and produce a flap lesion causing intermittent locking of the knee. In addition, the osteocartilaginous fragment that remains partly attached to the underlying bone can be revascularized and may heal with conservative therapy in the young patient with open growth cartilage. Either drilling to enhance vascularization or fragment fixation can be considered in older patients or if a conservative treatment fails. In case of total separation, the fragment floats free within the joint as a loose body, and surgical treatment becomes imperative.

MRI can detect the lesions early in the course of the disease, when there is separation of the first few layers of subchondral bone and cartilage from the underlying bone. Determination of the mechanical status of the fragment (loose or stable) is possible, and the progress of the lesion (nondisplaced, impending separation, complete displacement) can be followed. With three-dimensional imaging, the integrity of the hyaline cartilage overlying the bony anomaly can be clearly assessed in its whole thickness. The method is of great value, not only in planning the proper treatment but also as a follow-up tool, whether surgery has been performed or not.

MRI Appearance. The osteochondritis dissecans lesion appears as an area of low or moderate signal intensity on T_1 images, increasing on T_2-weighted images. Loosening is obvious when the fragment is displaced. If nondisplaced but sequestrated, the fragment is outlined by fluid, detectable at the interface with the parent bone as a line of high signal intensity on T_2-weighted images and moderate signal intensity on T_1-weighted images (Fig. 2.40). Lack of increase in signal intensity is seen in older lesions and can be explained by the presence of fibrosis (Mesgarzadeh et al. 1987; Mink et al. 1987).

2.3.5.2 Osgood-Schlatter Disease

Osgood-Schlatter disease is a partial avulsion of the tibial tuberosity at the lower insertion of the patellar tendon, occurring during adolescence growth (Fig. 2.28). Those presenting with Osgood-Schlatter disease are usually athletically active young teenagers, reporting anterior knee pain and tenderness located below the patella and increased by exertion (football, rugby, and jumping). In early stages, there may be no apparent radiological changes except swelling. Fragmentation of the epiphysis of the tibial tubercule appears in later stages. A comparable finding is described at the upper attachement of the patellar tendon in the Sinding-Larsen-Johansson syndrome. MRI can assess this disease at an early stage, evaluating not only the bone abnormalities as on conventional radiographs but also the precise alterations of the patellar tendon and the infrapatellar fat pad and joint recessus (Bodne et al. 1988).

2.3.5.3 Osteochondral Injuries

Combined lesions of bone and cartilage are quite common in the knee joint but are often incorrectly diagnosed initially (Fig. 2.38). They can be caused by impaction, avulsion, or shearing forces on the bone surfaces. Impaction of the

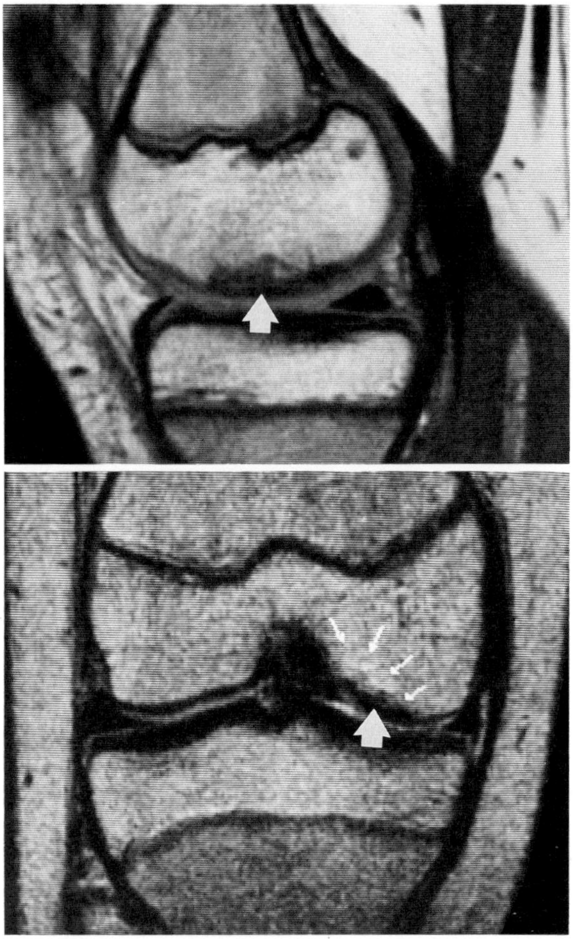

Fig. 2.40 a, b. Osteochondritis dissecans in an 11-year-old girl with unilateral knee pain. **a** Sagittal T_1-weighted image (SE 2200/22) shows a low intensity osteochondral lesion (*large arrow*). **b** The T_2-weighted image (SE 2200/80) displays a subchondral zone of low signal intensity (*large arrow*) surrounded by a crescent of high signal intensity (*small arrows*), partially due to edema. There was no displacement of the osteochondritic lesion

patella on the lateral femoral condyle, resulting from patellar dislocation and reduction, or combined torsional and compression forces on the flexed knee, in twisting injuries such as in football or skiing, are common traumatic mechanisms. They may occur anywhere in the knee but are most common in the lateral femoral condyle and the medial patellar facet (mainly in adolescents). The symptoms may mimic meniscal tears or more rarely ligamentous injuries. The prognosis of an osteochondral fracture is greatly dependent on its mechanical stability.

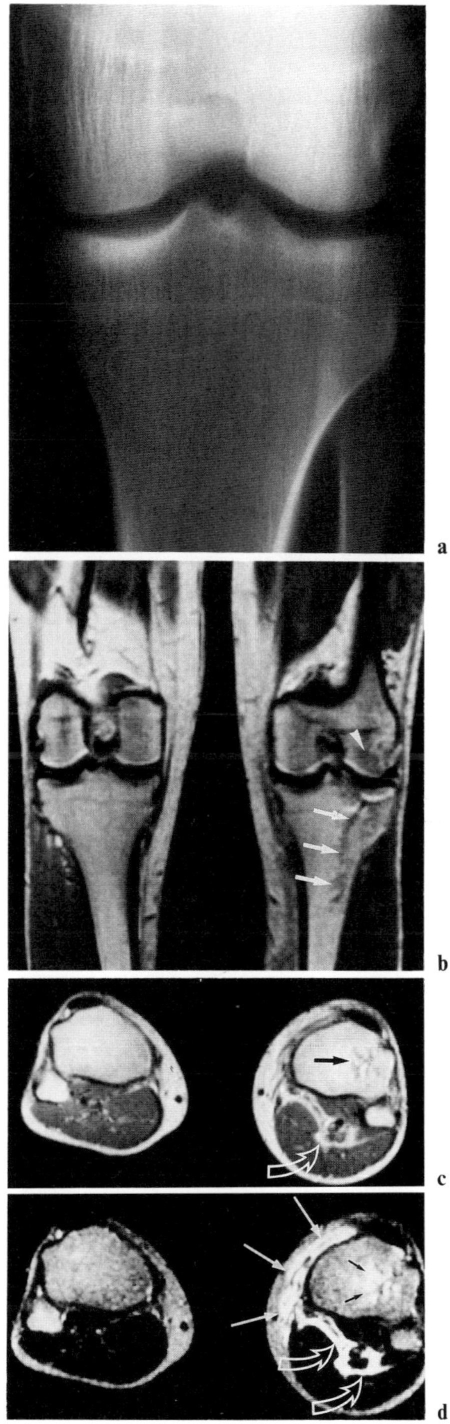

Fig. 2.41 a–d. Occult bone fracture in a 23-year-old motorcyclist admitted at the emergency department after a fall on the knee. Plain radiographs and tomograms (**a**) were negative, and the clinical examination suspected torn medial meniscus and collateral ligament. A preoperative MRI examination was performed and disclosed an occult fracture of the proximal lateral tibia, linear on coronal (**a**, SE 2000/22) and stellate on axial view (**b**, 2000/22; *arrows*). Associated bone contusion of the lateral femoral condyle was detected on the coronal section as a decrease in intramedullary signal intensity (*arrowhead*). No meniscal or ligamentous rupture was detected, and a conservative treatment was adopted. **c** On the axial T_2-weighted image (SE 2000/90), an anteromedial subcutaneous fluid infiltration (*thin arrows*) associated with an intermuscular collection (*open arrows*) were depicted with a high signal intensity. Note that the signal intensity of the tibial fractured region is diffusely brightened on T_2 (*black arrows*, **d**)

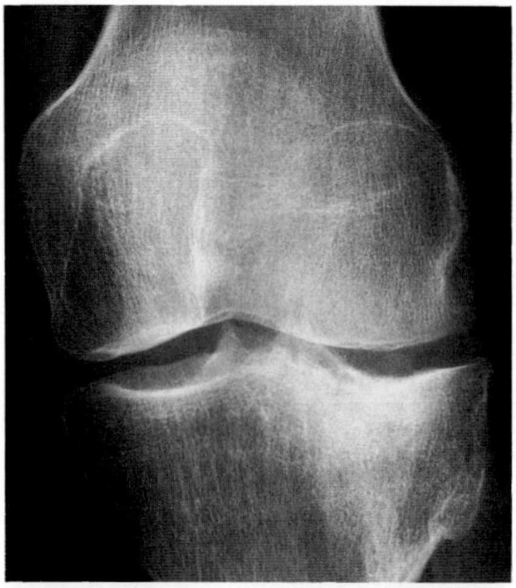
a

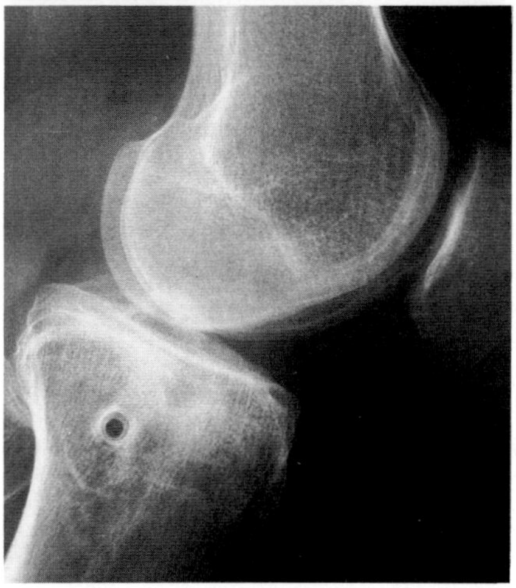
b

Fig. 2.42 a–e. Old fracture of the lateral tibial plateau, involving the articular surface in a 25-year-old judo brown-belt with persistent knee complaint. **a** Anteroposterior and lateral X-ray views (**b**) depict depression of the lateral tibial plateau with degenerative changes. Obliteration of the fracture line and presence of bone sclerosis are observed. On the coronal (**c**) and sagittal (**d**) proton-density images (SE 2500/22), the fracture looked healed (*large arrows*), while the tibial cortical bone shows irregularities of thickness and outline (*small arrows*). Intraarticular fragments of meniscal and cartilaginous origin (*arrowheads*) are floating free in the local joint effusion of intermediate signal on the T_1 weighted images and high on T_2 (**e**, SE 2500/80)

MRI Appearance. Displaced and impacted types are described. The articular cartilage and a thin segment of the underlying subchondral bone are sheared off and at least partially displaced. This is well depicted, particularly on three-dimensional images, where the cartilaginous faces are evaluated in their whole thickness. Partial attachment of the osteochondral fragment can be confirmed. Also, loose bodies are identified in the joint fluid or sometimes in paraarticular recessi. Impacted fractures are easily identified. In the subchondral bone, an area of low signal intensity is seen on T_1- and T_2-weighted images. This site is surrounded by a larger zone of edema, with higher signal intensity. The overlying cartilage presents irregularities of contour and signal inhomogeneities, depending on the importance of the fracture. Again, other associated traumatic lesions can be diagnosed simultaneously. MRI surpasses arth-

Fig. 2.42 c–e

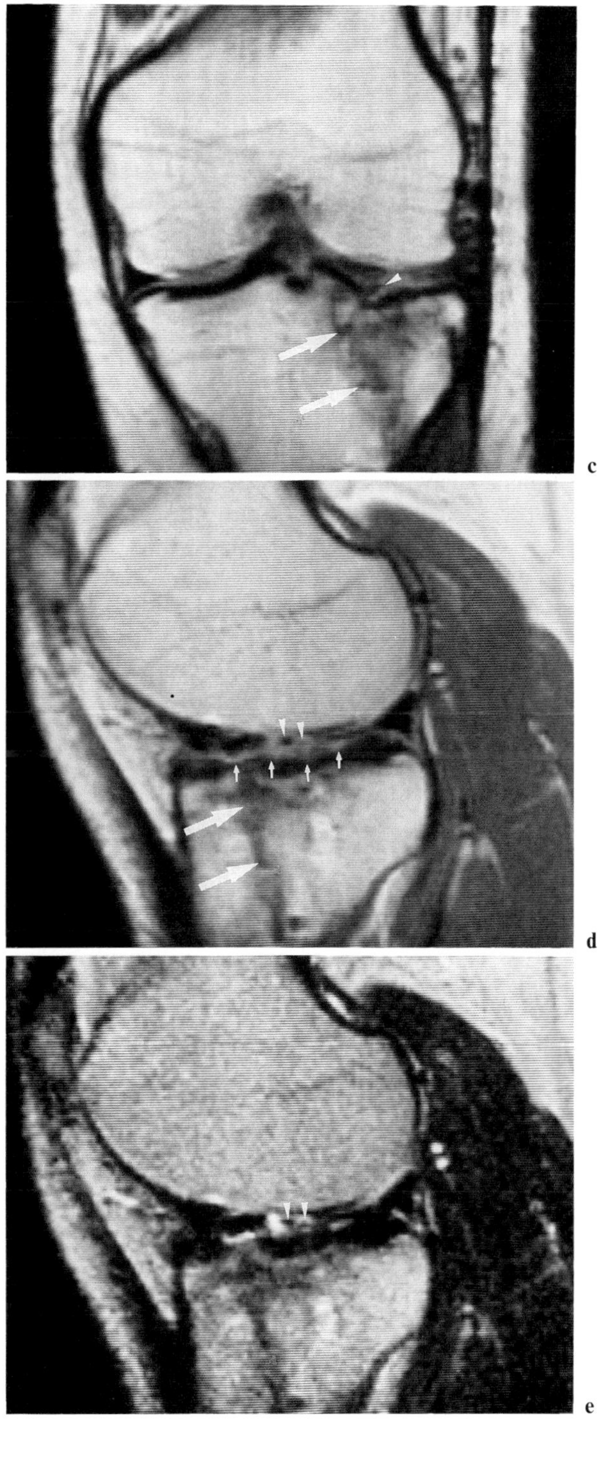

rography and arthroscopy, which can visualize only the superficial joint structures (Betran 1986 b; Mink 1987; Wojtys 1987; Mink and Deutsch 1989).

2.3.5.4 Occult Fractures

Sport-related trauma to the knee can produce strong clinical suspicion of fracture, while initial X-rays are normal (or equivocal). They are most commonly located in the (lateral) tibial plateau, resulting from stress phenomena or from valgus-type stresses associated with tears of the MCL (medial collateral ligament). Stress fractures of the knee occur in beginning joggers as well as in elite athletes in certain conditions (change of running surface, unusual physical activity, traumatic injury to another joint, etc.). They usually result from increased stress in normal bones and are also called fatigue fractures. They may localize in the posteromedial aspect of the proximal tibia in runners. They are classically described just below the medial tibial plateau in military recruits. Conventional radiographs are initially normal. Bone reactions become visible only a few weeks later. Bone scintigraphy results are usually positive, and this is a convenient tool for early diagnosis of stress fractures; however, it has a relatively low specificity and may pattern degenerative or inflammatory arthritis, synovitis, etc.

MRI can detect these fractures prior to their appearance on standard X-ray images and sometimes bone scans.

MRI Appearance. On T_1-weighted images they are seen in the medullary bone, as linear (or eventually globular) areas of very low signal intensity, surrounded by a poorly defined area with less dark signal intensity, which increases on T_2-weighted images. An extension into the adjacent cortical bone can be depicted. The fracture site may be more diffuse and amorphous, appearing as extensive areas of decreased signal on T_1-weighted images, without linear components. Occasionally, a thick band or multiple lines of low signal intensity on T_1-weighted images, brightening on T_2, are seen. Those characteristics help to distinguish it from a normal adult physeal line (Fig. 2.41). Apart from the analysis of the bone fracture, MRI allows the discovery of meniscal lesions, osteonecrosis or bursitis of the pes anserinus, which can mimic the initial symptoms of fatigue fractures (Stafford 1986; Lodwick 1987; Mink 1987; Lee and Yao 1988; Yao and Lee; Lynch 1989; Mink and Deutsch 1989).

2.3.5.5 Bone Contusion (Bone Bruise)

This bone abnormality is often detected after direct traumata and is associated with injuries of soft tissue structures, such as ACL.

MRI Appearance. A bone contusion appears as a poorly defined zone in the medullary bone situated under the cortex, characterized by a decreased signal intensity on T_2-weighted images. This is probably due to the presence of intraosseous edema, hyperemia, and hemorrhage (Mink and Deutsch 1989; Lynch et al. 1989; Yao and Lee 1988).

2.3.5.6 Tibial Plateau and Femoral Fractures of the Knee

Apart from the detection of radiographically nonvisible fractures, MRI allows an excellent evaluation of evident fractures. A topographic view of the fracture in the three orthogonal planes is afforded by MRI with determination of orientation and displacement of the fragments. The anatomic extent and configuration of the bone injury with degree of comminution and also depression of the tibial plateau can be demonstrated. With the improvement of image quality, evaluation of abnormalities of cortical bone is possible with good anatomic detail. Additional muscular, ligamentous, meniscal, and cartilaginous lesions are simultaneously diagnosed (Fig. 2.42).

MRI Appearance. Tibial fractures are often minimally displaced, vertically oriented, and extending into the plateaus. Femoral fractures are mostly Y-shaped, extending across the supracondylar region. Linear or serpiginous segments of diminished signal intensity are well delineated on T_1-weighted images. The fracture line is occasionally depicted with small step-off in articular tibial cartilage. Around the fracture, a broad area of decreased signal intensity on T_1-weighted images, due to edema and hemorrhage, is described. Usually, local edema is also present in the adjacent fat tissues. Older fractures have a low signal intensity on both T_1 and T_2 images. A disadvantage of MRI is the poor depiction of small calcifications, which yields low signal intensity. The noisy aspect of medullary bone on the fast three-dimensional images makes difficult the detection of medullary changes, which requires experience from the reader (Beltran 1986b; Mink 1987; Yao and Lee).

3 The Joints of the Ankle

P. VAES, M. SHAHABPOUR, and P.-P. CASTELEYN

3.1 Introduction

Injuries to the ankle and foot in sports are very frequent; 15%–25% of all sport injuries affect this part of the lower limb (Garrick 1988). It is surprising to see that 85% of these injuries are inversion distorsion traumata of the ankle (Nicholas and Hershman 1986). Ankle sprains account for 3%–10% of all emergency department visits (Brostrom 1965; Hume 1984; Muwanga 1986). The ankle joint is most vulnerable in sports such as volleyball, basketball, football, soccer, squash, badminton, cross country, gymnastics, and tennis (Tables 3.1, 3.2). Swimming and cycling can be advised as safe sports for the foot and ankle. In the United States, basketball and football are the most frequent causes of injury; in Europe, volleyball is most frequently mentioned. Injuries to the lateral collateral ligaments of the ankle are typically caused by the person being pushed out of balance while in suspension in the air, followed by landing on the lateral border of the foot, and the opponant's landing on one's foot in volleyball and basketball. Even in noncontact sports, sprains and overload injuries can affect the level of skill and participation of the athlete.

Table 3.1. Frequency of ankle injuries in high school. (From Garrick 1975)

Sport	Sex	Number of ankle injuries per 100 participants
Basketball	Male	13.1
	Female	11.4
Cross-country	Female	11.5
	Male	2.1
Football	Male	10.9
Badminton	Female	6.3
Track-and-field	Female	4.3
	Male	2.6
Gymnastics	Female	4.1
Soccer	Male	2.9
Baseball	Male	2.0
Volleyball	Female	1.1

Table 3.2. Frequency of ankle sprains in various sports by level of participation. (Raymakers 1979)

Sport	Frequency of ankle sprains	Percentage of participants in the sport
Soccer	43%	51%
Volleyball	20%	4%
Basketball	10%	1.5%
Gymnastics	15%	24%
Tennis	7%	15.5%

3.2 MR Anatomy

Coronal and sagittal anatomy of the ankle has been described using conventional tomography (Dihlmann 1982; Slafani 1983), and computed tomography offers the possibility of studying the soft tissues in oblique and axial planes (Lindjo 1979). These techniques are very adequate in illustrating bone, however their diagnostic value for soft tissue structures is limited. Clinically important structures such as the lateral collateral ligaments are inadequately shown by these methods. Stress radiography, when performed mechanically and bilaterally, has proven its value in identifying ligament ruptures (Eideken 1978; De Smet 1985), but these techniques are not always accurate.

Description of the nature and extent of pathological abnormalities by MRI must be developed. Because of limited experience there is the risk that normal anatomical variants are identified as being pathological. This may decrease the specificity of MRI and lead to overtreatment. In the ankle and foot MRI often depicts normal anatomical variants in bone and soft tissue structures (Noto 1989). Cortical bone may appear irregular and simulate osteonecrosis; the normal ligamentum talofibulare posterius (TFP) often has irregular edges that should not be misinterpreted as partial rupture; the normal attachment of the posterior tibial tendon on the os naviculare may present an heterogeneous signal; and the ligamentum deltoideum may appear as an inhomogeneous structure even when it is normal. In many cases we have seen fluid and edema in the uninjured talocrural joint, near and around the lateral malleolus; we have also seen edema in and around the peroneus brevis tendon in asymptomatic ankles.

3.2.1 Multiplanar Anatomy

Optimal surface coil images are achieved by positioning the talocrural joint in the center of the coil with the foot in neutral position (this is in 10° of plantar flexion). The foot and ankle are supported with soft pillows; the knee is in 20° of flexion and the hip in 10°–20° of internal rotation (so that the medial side of the

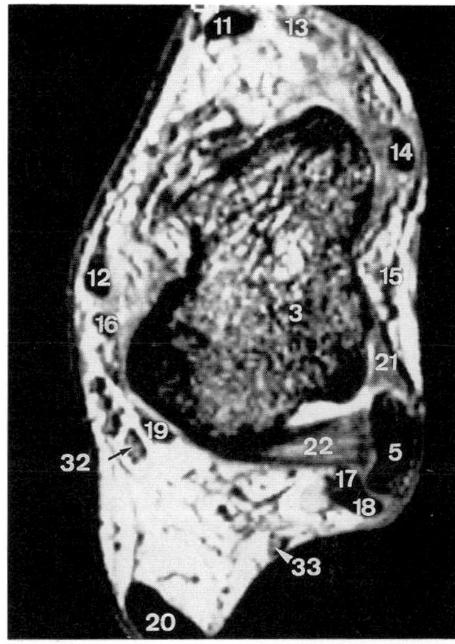

Fig. 3.1. Axial section through the distal tibiofibular syndesmosis (3 DFISP 40 30/10, 1 mm thick) where ligamentum talofibulare anterius and ligamentum talofibulare posterius are clearly visualized. *3*, Talus; *5*, lateral malleolus; *11*, tendon of tibialis anterior muscle; *12*, tendon of tibialis posterior muscle; *13*, tendon of extensor hallucis longus muscle; *14*, tendon of extensor digitorum longus muscle; *15*, tendon of peroneus tertius muscle; *16*, tendon of flexor digitorum longus muscle; *17*, tendon of peroneus brevis muscle; *18*, tendon of peroneus longus muscle; *19*, tendon of flexor hallucis longus muscle; *20*, Achilles tendon; *21*, anterior talofibular ligament; *22*, posterior talofibulare ligament; *32*, tibialis posterior artery and nerve; *33*, suralis nerve

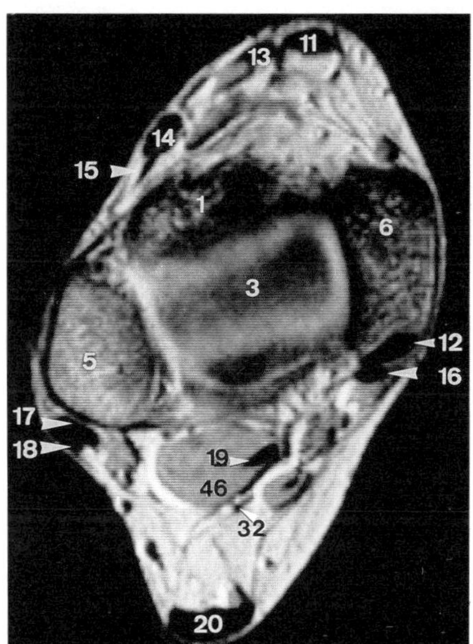

Fig. 3.2. Axial section through the talocrural joint (3 DFISP 40 30/10, 1 mm thick). *1*, Tibia; *3*, talus; *5*, lateral malleolus; *6*, medial malleolus; *11*, tendon of tibialis anterior muscle; *12*, tendon of tibialis posterior muscle; *13*, tendon of extensor hallucis longus muscle; *14*, tendon of extensor digitorum longus muscle; *15*, tendon of peroneus tertius muscle; *16*, tendon of flexor digitorum longus muscle; *17*, tendon of peroneus brevis muscle; *18*, tendon of peroneus longus muscle; *19*, tendon of flexor hallucis longus muscle; *20*, Achilles tendon; *32*, tibialis posterior artery and nerve; *46*, flexor hallucis longus muscle

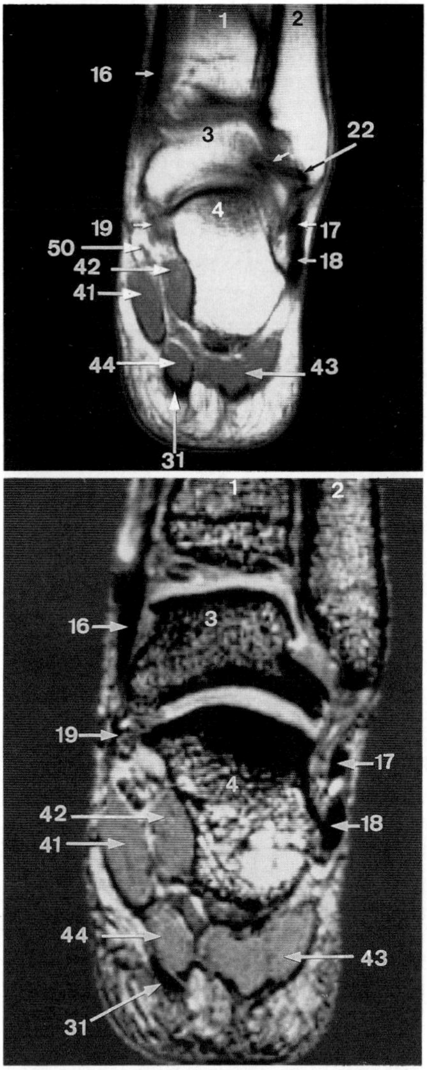

Fig. 3.3a, b. Coronal section of the joints of the ankle. **a** SE 2000/22. **b** 3 DFISP 40 30/10. *1*, Tibia; *2*, fibula; *3*, talus; *4*, calcaneus; *16*, tendon of flexor digitorum longus muscle; *17*, tendon of peroneus longus muscle; *18*, tendon of peroneus brevis muscle; *19*, tendon of flexor hallucis longus muscle; *22*, posterior talofibular ligament; *31*, plantar aponeurosis; *41*, abductor hallucis muscle; *42*, quadratus plantae muscle; *43*, abductor digiti minimi muscle; *44*, flexor digitorum brevis muscle; *50*, medial plantar neurovascular bundle

foot is in an exact vertical position); this is obtained by positioning a pillow under the hip and pelvis at the side of investigation. This neutral position of the lower leg is very comfortable and permits a decrease in motion artifacts. Reproduction of this standard procedure for three-dimensional analysis is easy to perform.

Standard imaging is performed in the three orthogonal planes. More specific imaging planes can be obtained using oblique spin-echo imaging or by a reconstruction system based on a three-dimensional sequence.

MRI in the axial plane provides clear depiction and localization of the flexor and extensor musculotendinous units (Fig. 3.1, 3.2). Their relationship

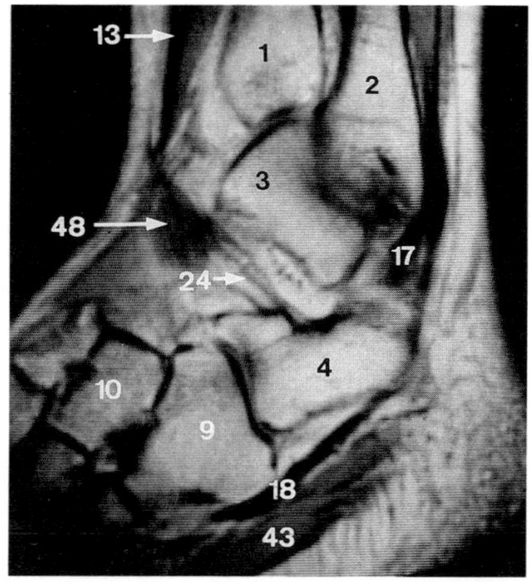

Fig. 3.4. Sagittal section through the lateral malleolus (SE 2000/22). *1*, Tibia; *2*, fibula; *3*, talus; *4*, calcaneus; *9*, cuboid; *10*, lateral cuneiform; *13*, tendon of extensor hallucis longus muscle; *17*, tendon of peroneus brevis muscle; *18*, tendon of peroneus longus muscle; *24*, interosseous talocalcaneal ligament; *43*, abductor digiti minimi muscle; *48*, extensor digitorum brevis muscle

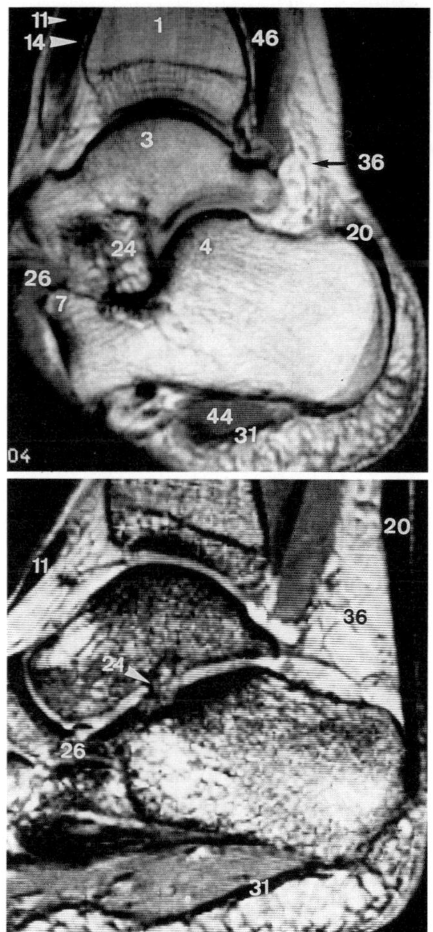

a

b

Fig. 3.5 a, b. Sagittal section through the subtalar joint with clear depiction of the ligamentum talocalcaneum interosseum. **a** SE 2000/22, 3 mm thick. **b** 3 DFISP 40, 0.8 mm thick. *1*, Tibia; *3*, talus; *4*, calcaneus; *7*, sustentaculum tali; *11*, tendon of tibialis anterior muscle; *14*, tendon of extensor digitorum longus muscle; *20*, Achilles tendon; *24*, interosseous talocalcaneal ligament; *26*, calcaneonavicular ligament (spring ligament); *31*, plantar aponeurosis; *36*, pre-Achilles fat pad; *44*, flexor digitorum brevis muscle; *46*, flexor hallucis longus muscle

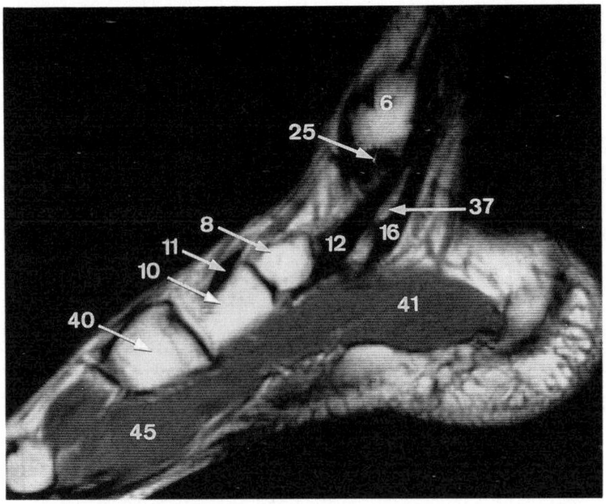

Fig. 3.6. Sagittal section through the medial malleolus (SE 2000/22, 3 mm thick). *6*, Medial malleolus; *8*, navicular; *10*, medial cuneiform; *11*, tendon of tibialis anterior muscle; *12*, tendon of tibialis posterior muscle; *16*, tendon of flexor digitorum longus muscle; *25*, deltoid ligament; *37*, medial tubercle of talus; *40*, first metatarsal; *41*, abductor hallucis muscle; *44*, flexor digitorum brevis muscle; *45*, flexor hallucis brevis muscle

to the tibia, fibula, talus, and calcaneus can be evaluated in every plane. The membrana interossea and parts of the anterior and posterior talofibular ligaments can be depicted. An image of plantar foot muscles is obtained in standard axial planes.

MRI in the coronal plane provides a sharp delineation of bone; a clear view of bone marrow and cortical bone is given (Fig. 3.3). The articular surfaces of the talocrural joint (particularly cartilage thickness) can be studied in detail on three-dimensional sequence. Hyaline cartilage has an intermediate signal intensity that distinguishes it from the low signal of subchondral cortical bone (Beltran 1986). On the medial side of the ankle, the different components of the ligamentum deltoideum can be recognized: pars tibiocalcanea, tibionavicularis, anterior and posterior tibiotalaris. On the lateral aspect, parts of the ligamentum calcaneofibulare (CF) can be visualized on standard coronal planes. On the medial side, the tendons of the flexor hallucis longus and flexor digitorum are depicted. The sinus tarsi and the ligamentum talocalcaneum interosseum can also be investigated on standard coronal views. The calcaneal tendon is partially seen on a few coronal images due to its curved trajectory.

MRI in the sagittal plane permits optimal assessment of the articular surfaces of the trochlea tali and the distal tibia (Figs. 3.4, 3.5, 3.6). Cartilage integrity can be assessed because of the relatively high signal intensity of hyaline cartilage compared to subchondral cortical bone (Beltran 1986). Normal ankles may show a small amount of fluid within the articular spaces. Differentiation of this fluid from the articular cartilage is difficult on T_1-weighted images. As fluid increases and cartilage decreases in signal intensity

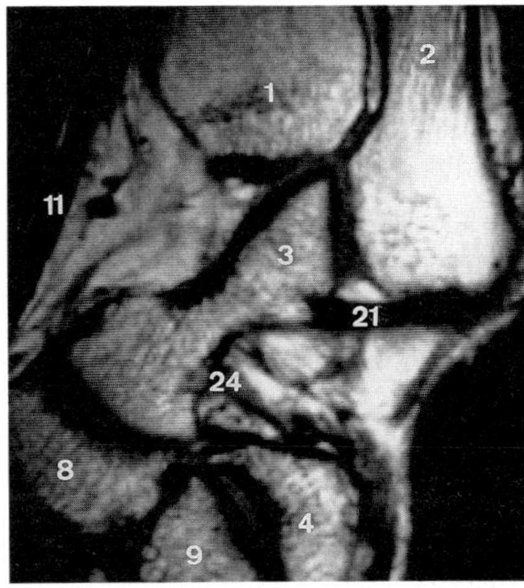

Fig. 3.7. Ligamentum talofibulare anterius as visualized in a sagittal oblique section (SE 2000/26). *1*, Tibia; *2*, fibula; *3*, talus; *4*, calcaneus; *8*, navicular; *9*, cuboid; *11*, tendon of tibialis anterior muscle; *21*, anterior talofibular ligament; *24*, interosseous talocalcaneal ligament

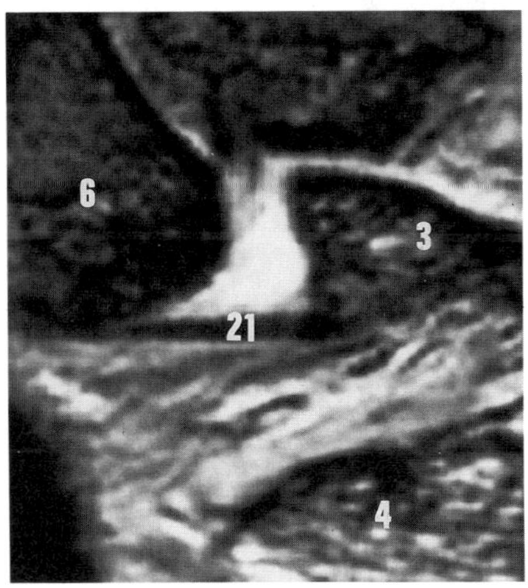

Fig. 3.8. Oblique reconstruction of the TFA ligament from axial 3DFISP40 (30/10). *3*, Talus; *4*, calcaneus; *6*, lateral malleolus (fibula); *21*, anterior talofibular ligament

on T_2-weighted images (for example TR of 2000 ms, TE of 70 ms), their distinction is feasible (arthrogram effect). However the precise delineation of the thin layers of hyaline cartilage is not possible in spin echo sequences (Hajek 1986). The interosseous talocalcaneal ligament can be seen on two subsequent 3-mm-thick standard sagittal planes. Depending on the ankle orientation, the first slice usually shows its proximal talar origin, and the second illustrates the attachment at the calcaneus.

The sagittal view offers the clinically most interesting image of the Achilles tendon. It can be studied down to its attachment on the calcaneus. More distally, the plantar fascia and the ligamenta plantaria are illustrated. Foot muscles and short flexors and extensors as well as abductors and adductors of

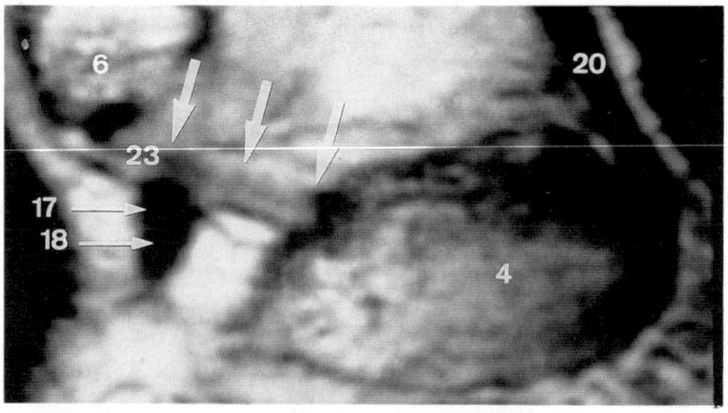

Fig. 3.9. Oblique reconstruction of the ligamentum calcaneofibulare from axial 3DFISP40 (30/10). *4*, Calcaneus; *6*, lateral malleolus; *17*, tendon of peroneus brevis; *18*, tendon of peroneus longus; *20*, Achilles tendon; *23*, calcaneofibular ligament

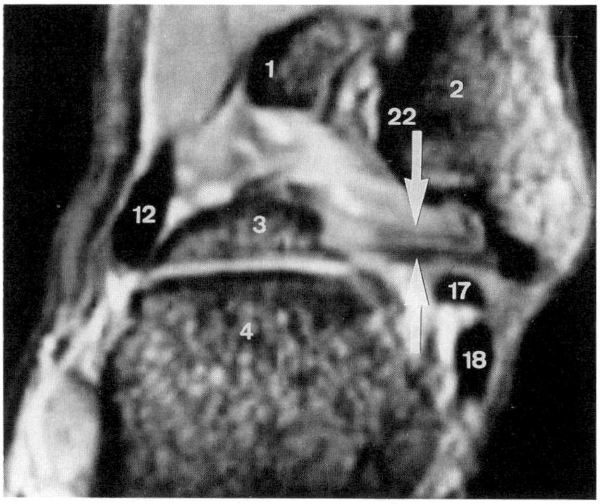

Fig. 3.10. "Coronal oblique" reconstruction of the ligamentum talofibulare posterius from axial 3DFISP40 (30/10). *1*, Tibia; *2*, fibula; *3*, talus; *4*, calcaneus; *12*, tendon of tibialis posterior muscle; *17*, tendon of peroneus brevis muscle; *18*, tendon of peroneus longus muscle; *22*, posterior talofibular ligament

the hallux can also be studied in detail. Evaluation of the normal anatomy of the lateral collateral ligaments of the ankle is possible by three-dimensional MRI. It is necessary to reconstruct oblique sections through the full length of the ligamentum talofibulare anterius (TFA), CF, and TFP to assess their continuity. With experience, it is possible to perform a series of oblique spin echo images through the lateral ligaments; the plane of acquisition is chosen on axial views (Fig. 3.7). There is always a risk that the structures are only partially visualized due to their anatomic variation. The advantage of three-dimensional MRI is the possibility to reconstruct oblique imaging planes at any time, after the data set of images is obtained. The 128 (0.7- to 1-mm-thick) slices of the single pulse sequence contain all necessary information.

The TFA ligament can be visualized in three to four subsequent 1 mm thick slices in the axial plane (Fig. 3.1). However, to obtain a full view of the length and width of the ligament, an oblique reconstruction is needed (Fig. 3.8). The CF ligament is much more difficult to depict, because of its subtle obliquity and higher signal intensity. There is lack of contrast between CF and the surrounding fat structures. Crossing the peroneal tendons, this ligament is not confined to one plane (Fig. 3.9). The TFP ligament is shown on axial and coronal slices (Figs. 3.1, 3.3). To depict it in full length and width oblique reconstructions through the short and long fibers are needed (Fig. 3.10).

3.3 Common Ankle and Foot Injuries

3.3.1 Ligamentous Disruption

Soft tissues, including ligaments, tendons, and cartilaginous articular surfaces of ankle and foot, are often damaged in acute trauma or inflammatory conditions. Clinical evaluation of the joint is followed by radiography; in some instances arthrography or tenography are performed. Multiplanar CT can also be used for evaluation of ankle injuries. Disadvantages of CT include radiation exposure, limitation of direct imaging to one or two orientation planes (usually axial or coronal, possibly sagittal plane), difficulties in positioning, and increased radiation dose when performing multiplanar images. In most cases, the combination of clinical and standard radiological findings provides enough information to evaluate the soft-tissue damages associated with ankle injuries. However, significant ligamentous and cartilaginous injuries may be overlooked.

In ankle sprain, accurate evaluation of the severity of injuries to the lateral collateral ligaments remains difficult. Pain, swelling, and functional impairment are unreliable clinical signs to support the diagnosis of ligament rupture; however, these signs influence the choice of therapy. A selection in the vast population of ankle distorsions should be performed with accuracy, based on clinical examination and mechanical stress radiological procedure, so that only 5% of the traumatic ankles are examined by MR imaging.

Personal experience is described here of MR imaging of the ankle following injury of the lateral collateral ankle ligaments, with special attention to depiction of normal anatomy and soft tissue lesions. (Patients were admitted to MRI study only after detailed clinical, standard, and stress radiological examination. Based on this standard procedure, among the 900 patients evaluated yearly only 50 are accepted for MRI.)

Use of T_2 SE and 3DFISP40 sequences improves detection of pathological areas. Small effusions become readily visible, although differentiating between posttraumatic, inflammatory, and infectious fluid collections still remains

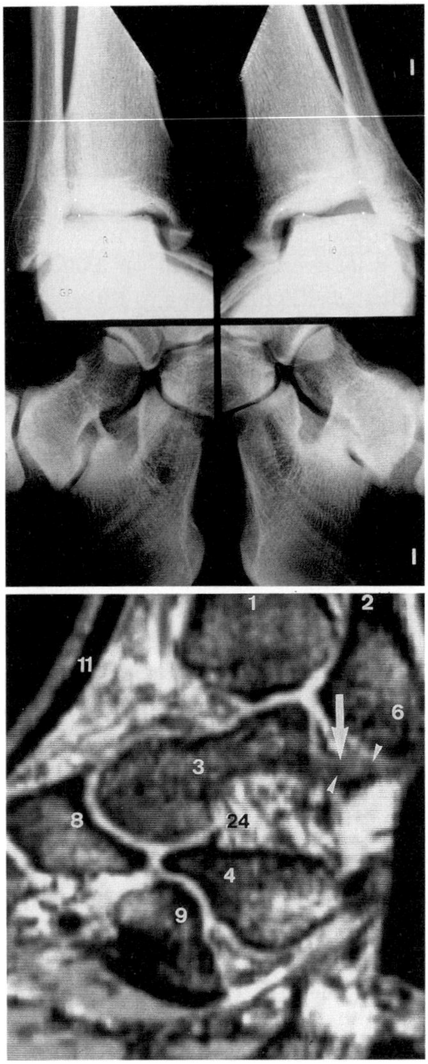

a

b

Fig. 3.11 a, b. Fibrotic ligamentum TFA in a 24-year-old female volleyball player presenting a chronic unstable talocrural joint. **a** Comparative anteroposterior and sagittal (anterior drawer) stress radiographs are performed. The left ankle (*l*) presents a positive (18°) talar tilt angle and a positive (11 mm) anterior drawer displacement. **b** "Sagittal oblique" view from an axial three-dimensional data set (30/10) shows a thickened anterior talofibular ligament (*arrow*) which presents an increased signal intensity with small areas of low signal (*arrowheads*) corresponding to fibrous tissue at surgery. *1*, Tibia; *2*, fibula; *3*, talus; *4*, calcaneus; *6*, lateral malleolus; *8*, navicular; *9*, cuboid; *11*, tendon of tibialis anterior muscle; *24*, interosseous talocalcaneal ligament

difficult. When we look at a chronic unstable ankle in which we can produce a talus rotation or talar tilt of more than 10°, we often find a fibrotic TFA (Fig. 3.11). A rupture of TFA ligament is not always recognized. As on spin echo oblique images, the ankle investigated with gradient echo sequences (3DFISP40), usually depicts the acutely TFA as an oedematous and vaguely visible structure (Fig. 3.12). If the ligament is elongated with partial rupture of the fibers, the ligamentous tissue is still seen as a continuous structure.

The appearance of normal as opposed to ruptured CF can be distinguished. While the intact ligament is seen as a continuous gray structure with well-delineated edges of low signal intensity, the ruptured ligament is difficult to

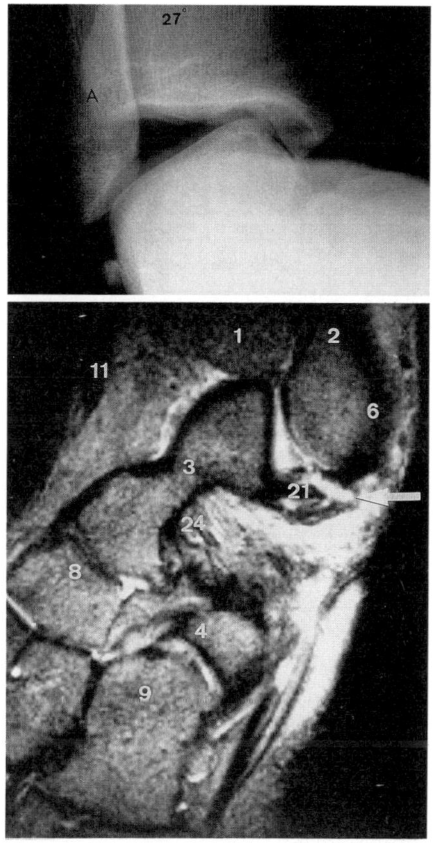

a

b

Fig. 3.12 a, b. Complete rupture of ligamentum TFA following an acute distorsion trauma. This 20-year-old basketball player performed a rebound and landed on an opponent's foot. **a** Anteroposterior stress radiograph shows a talar tilt angle of 27°. The patient reported immediate appearance of an egg-like swelling. Clinical examination showed positive anterior drawer sign and the presence of an important external hematoma associated with an internal malleolus contusion. **b** T_2-weighted "sagittal oblique" section (SE 2000/90) depicts disruption of the ligament fibers at the fibular point of attachment, where a local fluid collection is seen. The extensive associated subcutaneous fluid infiltration has a high signal intensity on T_2. *1*, Tibia; *2*, fibula; *3*, talus; *4*, calcaneus; *6*, lateral malleolus; *8*, navicular; *9*, cuboid; *11*, tendon of tibialis anterior muscle; *21*, anterior talofibulare ligament; *24*, interosseous talocalcaneal ligament

differentiate from the surrounding fat tissues. In chronic phase, MRI may help to define the sequels of ankle ligament injuries. These include recurrent instability (following severe distorsion trauma), sural nerve damage (following surgical repair), origin of recurrent sprains (fibrotic or elongated ligaments), recurrent peroneal dislocations, peroneal nerve palsy (cause of compression), and degenerative arthritis (Berquist 1989).

3.3.2 Tendon Ruptures

It is clear that MRI should not be used to establish the diagnosis of an Achilles tendon rupture. The appropriate clinical tests are extremely reliable and pathognomonic of this injury. However, if further information is wished regarding the condition, degree of separation, and orientation of the ends of the torn Achilles tendon, MRI is an accurate, noninvasive means of assessing these parameters (Keene 1989). MRI can help to evaluate the state (shredded, uniform, etc.) and orientation (antegrade, retrograde, etc.) of the torn fibers;

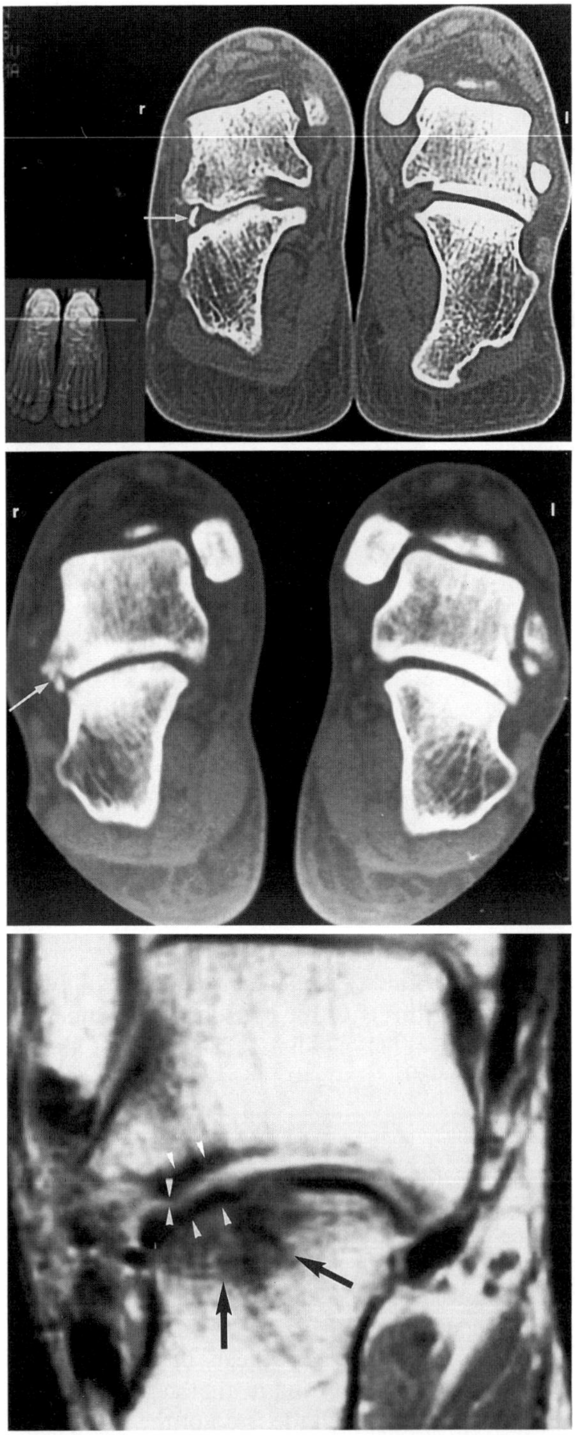

Fig. 3.13a–c

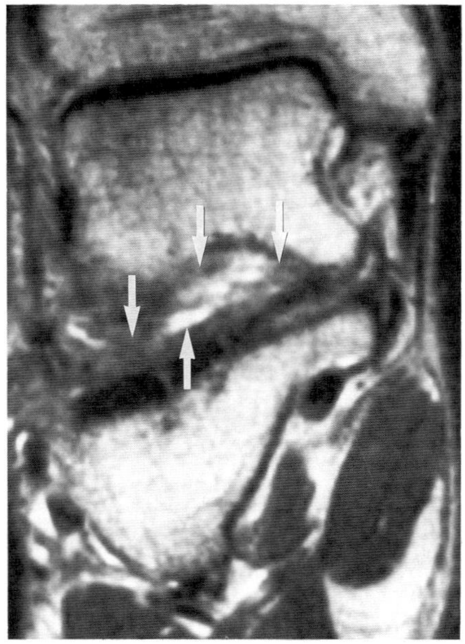

d

Fig. 3.13a–d. Detection of occult compression fracture in the subtalar joint combined with disruption of the talocalcaneal ligament. This 33-year-old athlete presented persistent pain, starting 1 year earlier, after a severe ankle injury. Bilateral coronal computed tomography, with bone (**a**) and soft tissue (**b**) windowing, shows an extraarticular avulsion fracture at the outer aspect of the right (*r*) subtalar joint (*arrow*). **c** Coronal MR view (SE 2400/22) discloses a broad area of intramedullary bone contusion in the calcaneum (*arrows*), associated with narrowing of the lateral subtalar space and sclerotic articular borders (*arrowheads*). **d** More anteriorly on coronal T_2-weighted image (2100/70), the interosseous talocalcaneal ligament is replaced by disorganized fibrous and edematous tissue, consistent with disruption of the ligament fibers (*arrows*)

the width of diastasis between the ends of the ruptured tendon can be assessed with or without ankle flexion. In acute complete disruption, the tendon ends are separated; a collection of fluid of high signal intensity is seen between the thickened tendon ends and in the tendon sheath. In partial disruption, the tendon does not look interrupted; a focal zone of high signal intensity is detected in the width of the tendon and in the tendon sheath (Keene 1986; Noto 1989; Berquist 1989).

3.3.3 Occult Bone and Cartilage Injuries of the Ankle

Bone contusion or bone bruise following severe ankle sprain is not rare at the talus; it is usually not visible on standard radiographs. Occult fractures can also be detected in the subtalar area (Fig. 3.13). Bone marrow edema of the talocrural bone structures is seen better on coronal or sagittal imaging planes. T_1-weighted images afford an obvious contrast between the low signal intensity of the lesion and the normal intense marrow fat. T_2-weighted images reveal typical increased signal of edema.

3.3.4 Stress Fractures

Stress fractures occur after repetitive minor trauma that is insufficient to cause an acute fracture. Most stress fractures are caused by excessive muscular activity rather than by direct trauma. In the lower leg, stress fractures are usually

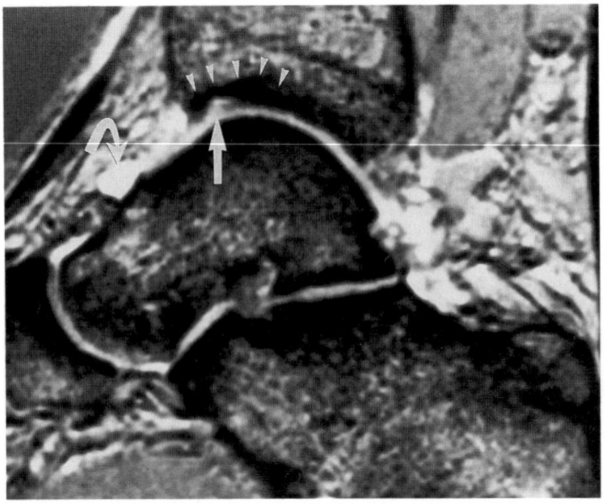

Fig. 3.14. Posttraumatic tibial osteochondral lesion in a 19-year-old hockey player who was hit by a hockey stick on the anterior part of the ankle. On sagittal 1-mm-thick section (3DFISP 40 30/10), fluid is distending the anterior part of the talocrural joint recessus (*curved arrow*). There is subchondral sclerosis (*arrowheads*) pointing out a subjacent cartilage alteration (*straight arrow*). Arthroscopy disclosed a cartilage fissure of the tibial articular surface

fatigue fractures caused by abnormal load on normal bone (Daffner 1987). Over 80% of adult stress fractures involve tibia, fibula, metatarsal bones, and calcaneus (Wilson 1969; Meurmann 1981; Berquist 1989; Resnick 1989). Clinically, a stress fracture presents as a tender area, sometimes with slight swelling and local temperature elevation. They can occur in the distal tibia in sports such as running, marching, and ballet; stress fractures of the distal fibula are described in runners and parachute jumpers. Metatarsal stress fractures are often associated with marching, running, and ballet. Tarsal location is less common (long distance running, marching); stress fractures of the navicular bones are described in basketball injuries and in running, and calcaneal fractures in jumping and running. The sesamoids can break in marching, skiing, and cycling.

Two sites of talar fractures that are relatively frequent in sport injuries and are often difficult to detect by standard radiography are the lateral and posterolateral processes. The latter may cause persistent posterior ankle pain and should not be confused with the sesamoid os trigonum.

MRI is indicated for early detection of stress fractures when they cannot be visualized on standard radiographs. ^{99m}Tc-labeled bone scans show increased uptake in the pathological area. MRI may detect the stress fracture at an earlier stage but also helps to locate precisely the area of increased uptake and to define its nature.

Cartilage damage may occur in ankle trauma when the joint is injured by a compression force combined with a rotational or gliding force. Articular

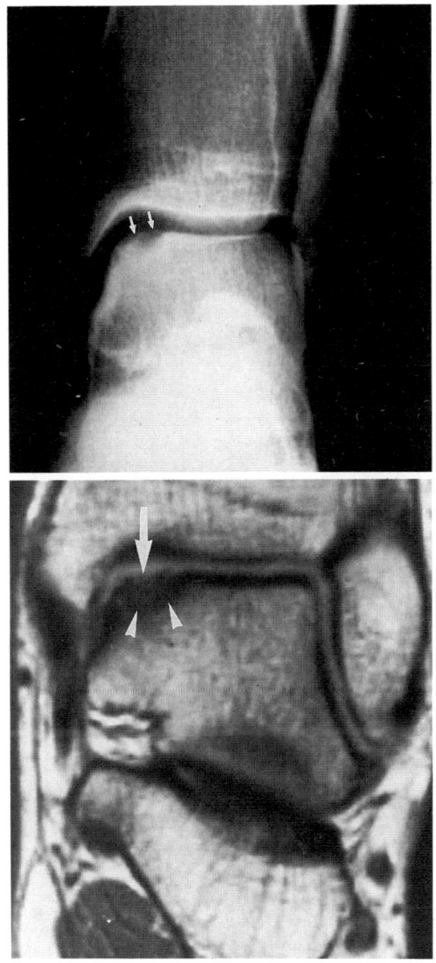

a

b

Fig. 3.15 a, b. Osteochondral lesion of talus in a 25-year-old baseball player who sustained an inversion ankle trauma 8 months earlier and complained of persistent pain. **a** Anteroposterior radiograph shows a bone defect at the medial talar dome (*small arrows*). **b** MR coronal view (2300/22) discloses presence of osteochondral fragments, emitting a hypointense signal (*large arrow*). The lesion is bordered by a sclerotic rim of low signal (*arrowheads*) and is not displaced. Disruption of the superior subchondral bone is depicted

cartilage of the talus may be severely damaged. Chondral lesions are sometimes associated with injury of subchondral bone (osteochondral lesion; Fig. 3.14). Osteochondral lesions of the talar dome include osteochondral fractures and osteochondritis dissecans. The advantage of three-dimensional imaging becomes especially clear in determining the status of the articular cartilage and the defect in the underlying bone (stability, attachment or complete/incomplete separation, displacement of the osteochondral fragment; Fig. 3.15). Presence and location of intraarticular loose bodies can be assessed (Yulish 1988), probably better with three-dimensional imaging. Such information is valuable in management of osteochondral lesions (preoperative planning).

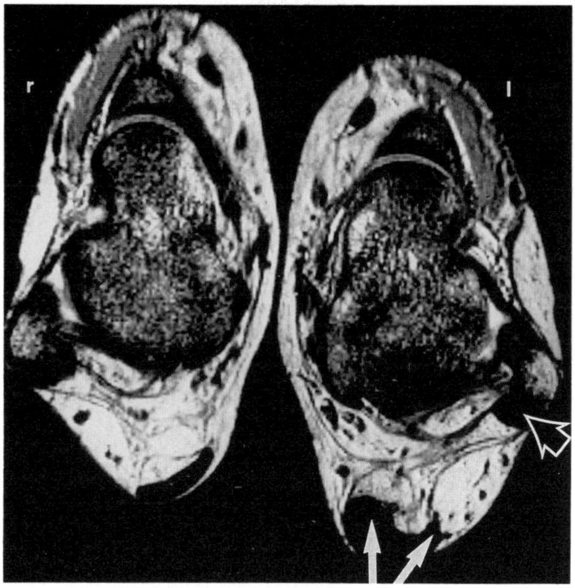

Fig. 3.16. A 22-year-old female long-distance runner presenting recurrent tendonitis of the peroneal and Achilles tendons. On the comparative 1-mm-thick axial view (3DFISP40 30/10), the left (*l*) ankle clearly shows major thickening of the Achilles tendon (*straight arrows*) and of the peroneal tendons dorsal to the lateral malleolus (*open arrow*)

3.4 Chronic Disorders

3.4.1 Tendonitis and Tenosynovitis

The tendons that cross the ankle include: (a) the tendo peroneus longus and brevis (lateral); (b) the Achilles tendon (posterior); (c) the tendo tibialis posterior, flexor digitorum longus, and flexor hallucis longus (medial); and (d) the tendo tibialis anterior, extensor hallucis longus, extensor digitorum longus (four tendons), and peroneus tertius (anterior). All tendons are enclosed in sheaths with the exception of the Achilles tendon.

Overuse can cause swelling and inflammation of the tendon tissue. The tendons may dislocate following inversion trauma. They may also rupture after repetitive injections of steroids (local injection or systemic administration) or in chronic inflammatory diseases (Morti 1977; Gilula 1984; Rosenberg 1987).

The peroneus brevis tendon is seen in the axial plane under the lateral malleolus. The peroneus longus tendon is localized beneath and more laterally, varying with the level of the image. The two tendons are separated by fat tissue of high signal intensity. Tendonitis and tenosynovitis are clearly depicted by means of MRI. The signal intensity appears brighter within the tendon in tendinitis. In chronic phase, the tendon may appear thickened with low signal

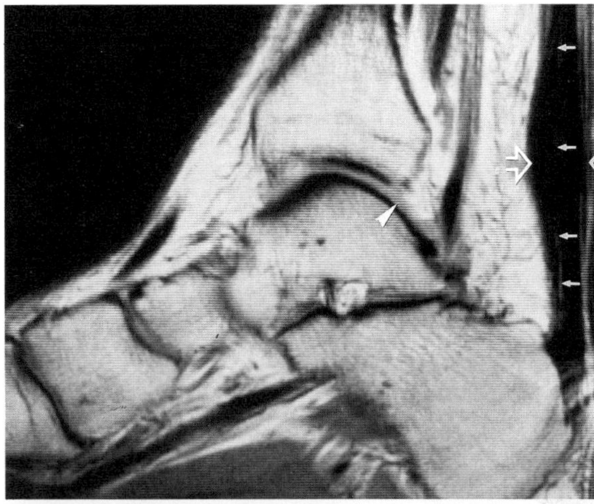

a

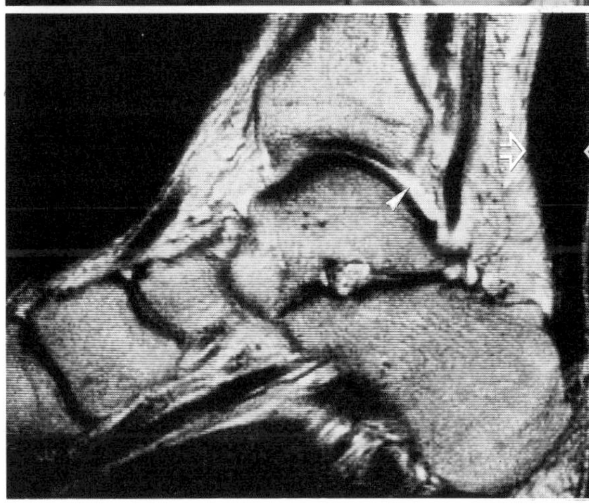

b

Fig. 3.17 a, b. Chronic Achilles tendinitis in a 39-year-old volleyball referee with intermittent pain and tenderness at the Achilles tendon area. Sagittal 3-mm-thick sections (SE 2500/22, 70) display extensive fusiform thickening of the Achilles tendon (*open arrows*). Thin linear areas of higher signal intensity (*small arrows*) are detected on proton-density image, consistent with small fissures in the tendon. Slight tibiotalar effusion is seen, having intermediate signal intensity on T_1-weighted image, high on T_2 (*arrowheads*)

intensity (Fig. 3.16). In tenosynovitis, the tendon is surrounded by fluid, which distends the tendon sheath.

The tendon fibers of the Achilles tendon cross approximately 2–6 cm above its calcaneal insertion. The posterior fibers course medially to laterally, and the anterior fibers cross laterally to medially prior to inserting in the calcaneus. Most tendon tears occur just above this area, perhaps owing partially to the reduced blood supply in this region. The Achilles tendon does not have a true tendon sheath (Kelikian 1985).

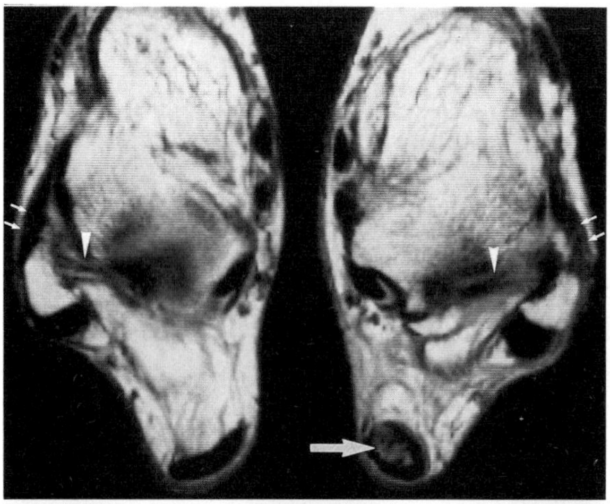

a

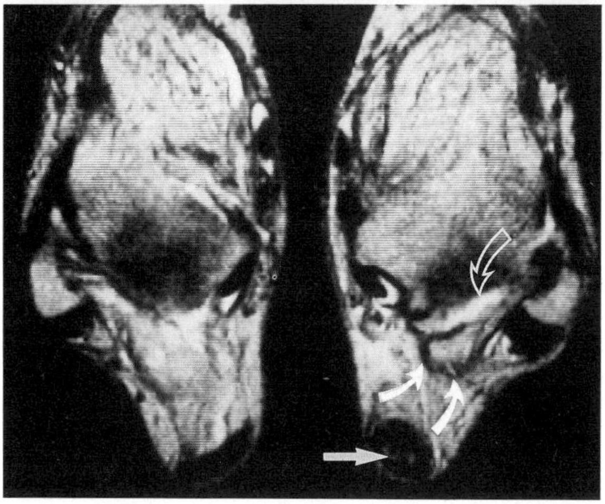

b

Fig. 3.18 a, b. A 20-year-old badminton player presenting chronic retrocalcaneal pain due to Achilles tendonitis. On bilateral axial 3-mm-thick sections (SE 2300/22, 70), there is a round-shaped swelling of the left Achilles tendon with an ill-defined central zone of high signal intensity (broader on proton density), consistent with necrotic tissue at surgery (*arrow*). Slight infiltration of the fat tissues by fluid of high signal intensity on T_2 (*open arrow*) and fibrosis (*curved arrows*) are depicted on the pathological side. Note the normal bilateral TFA (*small arrows*) and TFP (*arrowheads*) ligaments

MRI is especially useful in detecting and describing complete, partial, and old Achilles tendon injuries. On axial images, inflamed tendons may present higher signal intensity compared with adjacent normal tendons. In chronic inflammation, the tendons appear as thickened foci of low signal intensity (Fig. 3.16). In tenosynovitis, the foci of low signal intensity are surrounded by a high signal intensity fluid layer (target sign). Chronically torn tendons appear thickened or partially interrupted (Keene 1986; Noto 1989; Berquist 1989). A

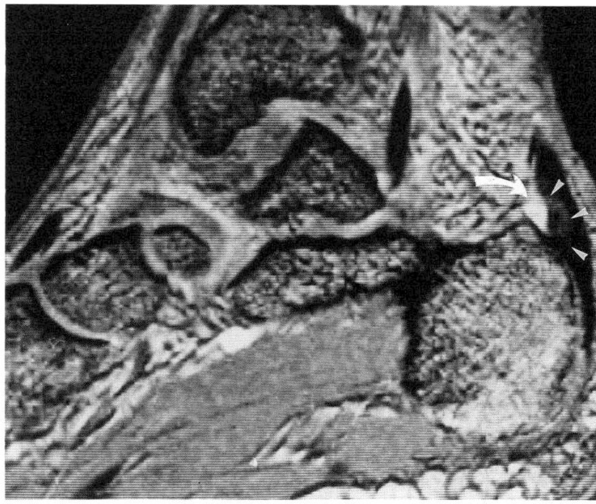

Fig. 3.19. Retrocalcaneal insertion tendinitis and bursitis in a 25-year-old gymnast due to overstrain. Sagittal 1-mm-thick slice (3DFISP40 30/10) shows a focal increase of signal intensity within the deep fibers of the calcaneal attachment of the Achilles tendon, probably corresponding to partial microtearing (*arrowheads*). Local fluid collection is seen in the pre-Achillic bursa (*curved arrow*)

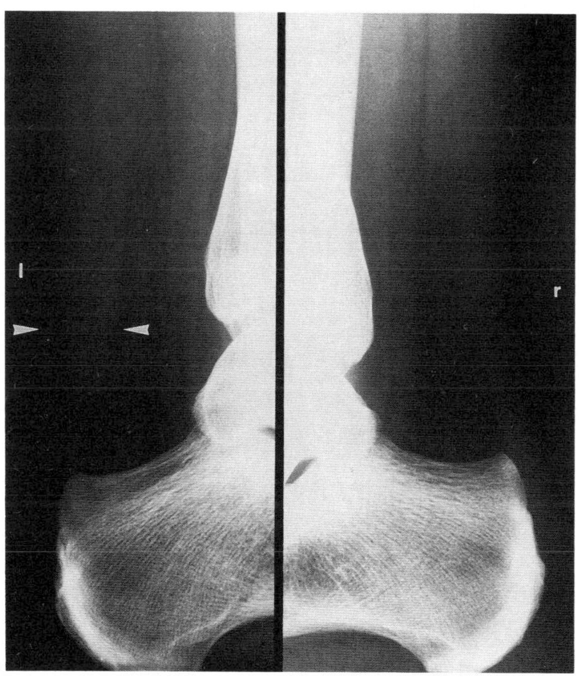

a b

Fig. 3.20 a–b (legend see p. 94)

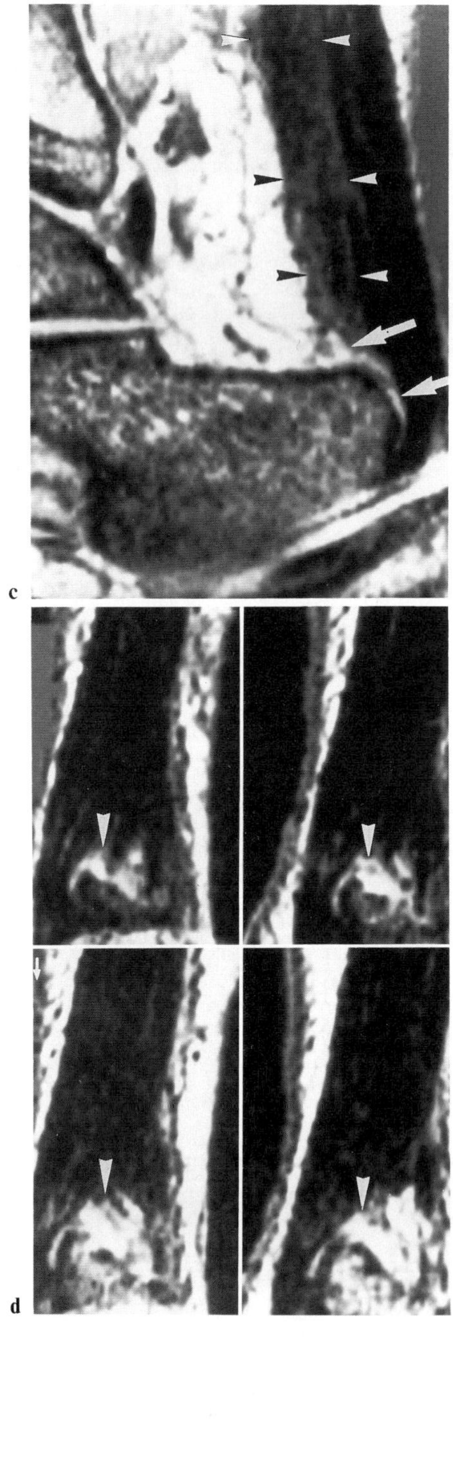

Fig. 3.20 a–d. Postoperative changes of the Achilles tendon. Tendonitis and fibrosis are displayed in a 23-year-old female marathon runner having persistent swelling and pain with a history of tendon shaving for recurrent tendonitis. **a, b** Conventional lateral X-rays show on the left (*l*) significant thickening of the whole tendon (*arrowheads*) with wavy and irregular anterior margin. Compared with the opposite side, the fatty Kager's triangle has a slight increased opacity due to edema. **c** Sagittal reconstruction from a three-dimensional axial data set (3 DFISP 40 30/10 1 mm thick) depicts two layers in the thickened tendon. The deep layer has an increased and heterogeneous signal intensity, consistent with fibrotic and edematous infiltration (*arrowheads*). Adjacent distension of the retrocalcaneal bursa by a small fluid collection of high signal intensity is also detected (*arrows*). **d** Coronal contiguous 1-mm-thick reconstructed sections show a round area with irregular borders at the tendinous insertion on the calcaneum (*arrowheads*) corresponding to focal necrosis

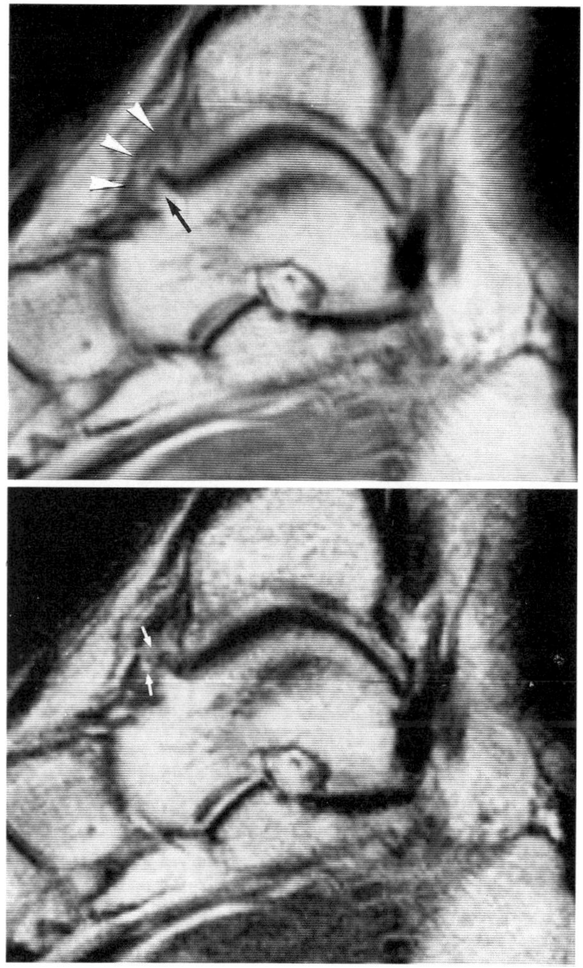

a

b

Fig. 3.21 a, b. Anterior capsular tear in a 33-year-old triple-jumper complaining of recurrent anterior ankle pain. **a** On 3-mm sagittal section (SE 2300/22), a sharp traction-type osteophyte of the neck of the talus (*black arrow*) is observed, associated with avulsion injuries of the overlying anterior capsule (*arrowheads*). **b** Fluid and edema infiltrate the capsular fibers on T_2-weighted image (SE 2300/70) (*small arrows*)

slight increase in signal intensity is depicted in the area of tearing; this is better depicted on proton density images. 3 DFISP sequences are the most sensitive in detecting microtears of the tendons. Evaluation of bursitis and other causes of Achilles tendon pathology is allowed as well. T_2 and especially three-dimensional sequences (3 DFISP 40) can be used to assess severity and extent of tendon necrosis, preoperatively and in postoperative follow-up (Figs. 3.17–3.20).

Of the medial tendons (posterior tibial, flexor digitorum longus, and flexor hallucis longus), the posterior tibial tendon, which is the most anteriorly

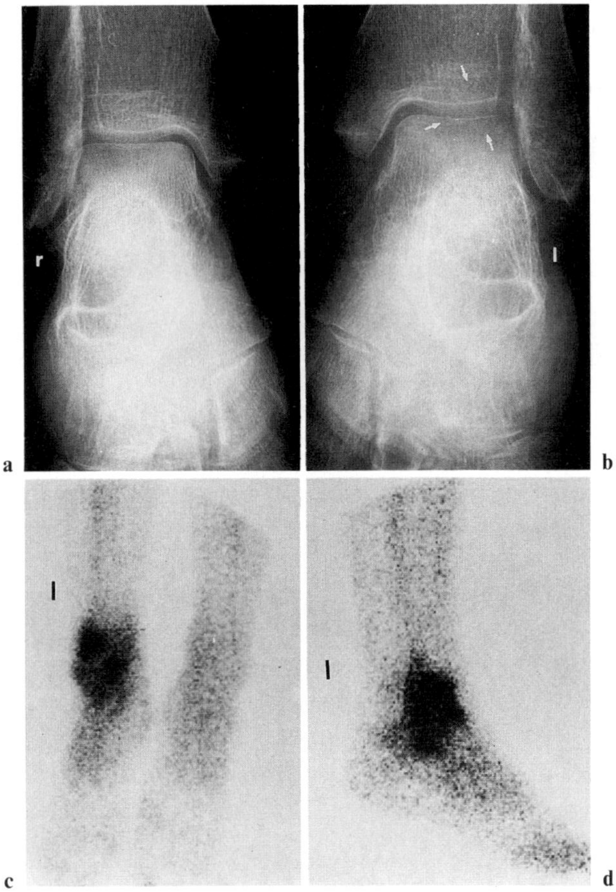

Fig. 3.22 a–e. Reflex sympathetic dystrophy syndrome in a patient with swelling of the ankle and refractory pain, after minor ankle injury. **a, b** Standard radiographs show slight regional bone demineralization and thinning of cortical bone in the left (*l*) joint (*arrows*), associated with mild periarticular soft tissue swelling, **c, d** Bone scanning detects a markedly increased uptake at the left ankle. **e** T_2-weighted MR coronal section (SE 2000/80) displays "mottled" inhomogeneity of signal intensity of tibia and talus, with multiple small areas of intense signal (*arrows*). This feature is not specific for reflex sympathetic dystrophy (but do not suggest aseptic necrosis)

located, is the most commonly injured (Kelikian 1985). Tendonitis of this tendon is often seen in the athlete with pes planovalgus. Rupture of the posterior tibial tendon may lead to progressive flat foot deformity. Nature and severity of the tendinous lesions can be assessed by MRI.

The anterior tendons are the anterior tibial, extensor hallucis longus, and extensor digitorum longus. In long-distance runners, injury to the anterior tibial tendon is regularly observed. On axial and sagittal MR images, the high signal intensity of fluid around or within the tendon tissue provides excellent

Fig. 3.22e

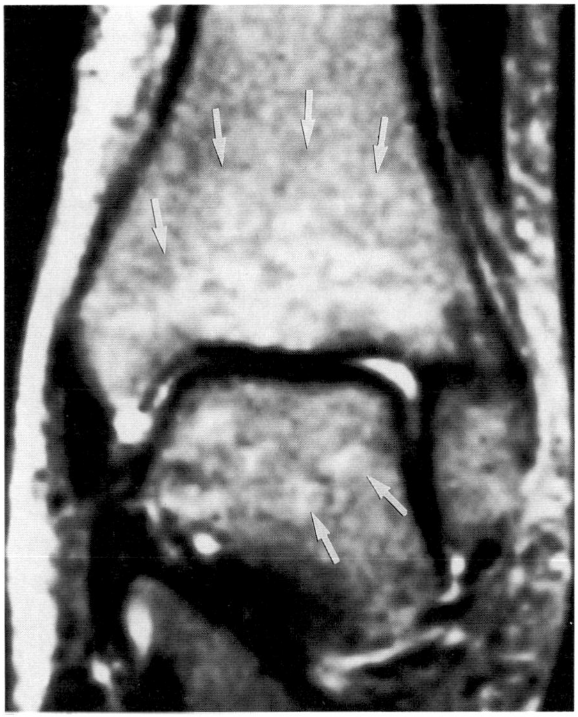
e

contrast between interstitial tissue and tendon of low signal intensity, especially on T_2-weighted sequences. Three-dimensional sequences allow a better detection of slight edematous changes and microruptures.

Plantar fasciitis occurs in more than 7% of runners. Reactive inflammation to overuse occurs to the plantar fascia insertion at the medial calcaneal tubercle. Focal inflammation in the middle of the fascia fibers may be seen as well. Detection of plantar fasciitis is possible at an early stage using three-phase ^{99m}Tc-labeled scans or MRI performed in the sagittal plane (Berquist 1989). Calcaneal spur is the result of mechanical stress acting through the plantar fascia into its origin at the medial calcaneal tubercle. Initially, chronic plantar fascia strains lead to inflammation within the plantar fascia. This stimulates the build-up of new connective tissue. Irritation of the periosteum is followed by inflammatory erosive changes and proliferation of new bone: a traction spur is formed. Pain is localized over the medial calcaneal tubercle, more posteriorly than for the plantar fasciitis.

Chronic capsular tears may occur in soccer players, recurrent ankle sprains causing avulsion injuries of the anterior capsule. The repeated minor injuries produce traction-type osteophytes at the neck of the talus. The torn capsular fibers can be shown on sagittal images (Fig. 3.21).

Foreign bodies in soft tissues of the foot are difficult to identify on X-rays, if not radiopaque. MRI can detect them more easily; thanks to the multiplanar

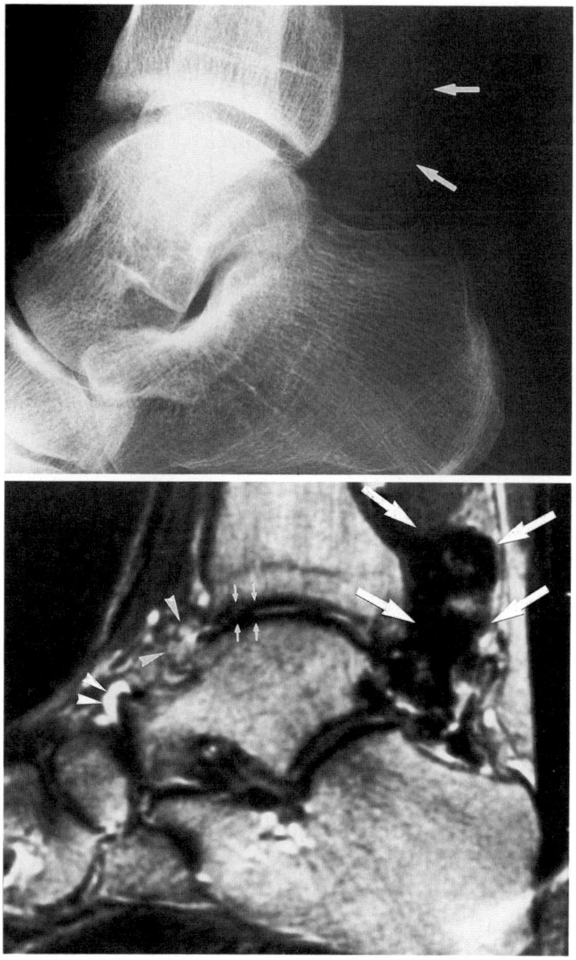

a b

Fig. 3.23 a, b. Synovitis in a 40-year-old amateur soccer player with history of long-term ankle pain. **a** Lateral X-ray view shows a soft tissue mass (*arrows*), filling Kager's space. Bone scintigraphy is clearly positive at the talocrural joint, but not specific. **b** MR sagittal image (2100/70) depicts a low signal intensity mass in the posterior paraarticular recessus. Presence of synovial proliferation with hemosiderin deposition (responsbile for the very shortened T_2) are characteristic of pigmented villonodular synovitis (*arrows*). Note also a moderate distension of the anterior joint recessi (*arrowheads*) and tibiotalar degenerative lesions (joint space narrowing and sclerotic margins; *small arrows*)

imaging possibilities it can clearly define location and relationship to the surrounding soft tissues (Berquist 1989). The foreign body can be seen as a small area of signal loss corresponding to its shape.

The tarsal tunnel syndrome is a rare affecion consisting in entrapment of the posterior tibial nerve as may follow an injury. It is localized medially, but pain may occur diffusely on the plantar foot and may radiate distally with tingling, burning, and numbness. MRI axial and oblique reconstructions based on three-dimensional imaging may visualize the affected region. MRI can help to identify the contributing factors to compression, for example, old fracture, midtarsal osteophytes, fibrous tissue deposits, chronic tendinitis, and tumoral mass. The nervous structures can be compared with the contralateral side, namely in case of doubtful electromyographic study.

The structures of the hindfoot, and particularly the subtalar region, are not easy to evaluate by clinical examination and standard imaging procedures. CT scan has shown interesting applications in the detection of sport-related bone and soft tissue lesions. Little experience has been accumulated with MRI. However, this method presents promising advantages. Occult fractures of the subtalar bones have previously been described (Sect. 3.3.3). The interosseous talocalcaneal ligament can be visualized in coronal and axial spin-echo sequences (Fig. 3.13). The thin three-dimensional images can better depict the components of the ligament as well as the thin cartilaginous layers of the subtalar joint. Tarsal coalition, most often seen in the calcaneonavicular and the middle facet of the talocalcaneal joint are easily diagnosed as with CT (Mosier 1984). On axial T_1-weighted images, bone marrow is present in the articular region, disclosing bony coalition. In case of fibrous or cartilaginous coalition, the signal intensity is low (Berquist 1989). MRI provides a topographic view of the abnormalities in all orthogonal planes.

3.4.2 Reflex Sympathetic Dystrophy Syndrome

This entity has been reported as a complication of trauma, surgery, and infection. Patients may complain of severe pain, tenderness, and stiffness. Because of the modification of water content in bone marrow, MRI seems to be a sensitive method for detection of this affection. A diffuse regional decrease in the signal intensity of bone marrow is observed on T_1 sequences. T_2-weighted images display a mottled inhomogeneity of signal intensity, with multiple small areas of intense signal. These findings are not specific to the reflex sympathetic syndrome but help to differentiate against a vascular necrosis, which has a more sectorized presentation (Fig. 3.22).

In inflammatory osteoarthritis, early modification of water content in bone marrow can be detected as well as joint effusion. On T_1 images, signal intensity decreases, while it brightens on T_2-weighted images. Pigmented villonodular synovitis is a monoarticular synovial disease of unknown origin that can affect the ankle joint and tendon sheaths in young adults. Synovium cannot be directly visualized by standard radiographs and arthrography. It presents typical characteristics on MR imaging. T_1 and T_2 images depict well-delineated nodular synovial masses presenting a low signal intensity due to the paramagnetic effect of hemosiderin (Fig. 3.23).

4 Hips and Pelvis

M. Shahabpour, C. Pierre-jerome, and K. Vandenborne

4.1 Introduction

The hip is an intrinsically stable ball-and-socket joint encased in a thick capsule and surrounded by large muscles protecting it from injuries. Fractures or dislocations are rare in nonvehicular sports due to the lower energy level. These usually occur in motorized sports or in high-risk or high-speed activities such as hang-gliding and downhill skiing. The hip joint is more often the site of stress fractures or spontaneously occurring inflammatory lesions which may be revealed in overuse conditions.

MRI is an excellent problem-solving tool for the assessment of a great variety of common sport-induced injuries. In professional athletics, for practical purposes, thorough evaluation and precise diagnosis of these lesions are extremely important. Adequate therapy is necessary to avoid unfortunate situations such as long term-disability.

The primary approach to the diagnosis of most sport-related injuries is still standard radiography. However MRI has an expanding role in the detection and evaluation of many sport-related soft-tissue and bony lesions. Here we describe the MR appearance of these lesions, which are difficult to identify on plain radiographs. The pathologies which are relevant in such context are avascular necrosis, reflex sympathetic dystrophy, and microtraumatic lesions.

4.2 MR Anatomy of the Hip

The subject can be divided into two domains: (a) bone structures (femoral head and neck, acetabulum) and (b) intra and extraarticular coxofemoral structures capsule, tendons, and ligaments; acetabular labrum; articular cartilage; muscle groups; neurovascular structures.

The normal bone marrow of the femoral head and neck has a high signal intensity and is surrounded by thin cortical bone of low signal intensity (Figs. 4.1, 4.2). A thinner curvilinear line of low signal intensity, representing the growth plate remnant and the fovea capitis should not be confused with pathological findings. The fovea capitis appears as an irregular depression located medially and is the point of attachment of the ligamentum teres. In the

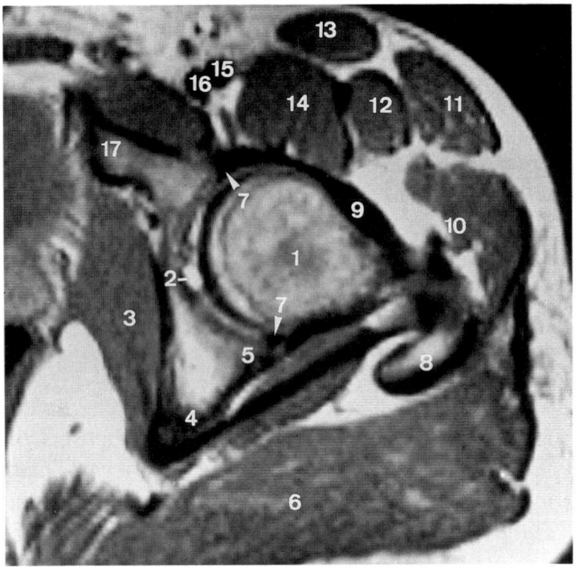

Fig. 4.1. Axial section (SE 2100/22) in a 15-year-old boy. *1*, Head of femur; *2*, acetabular fossa; *3*, internal obturator muscle; *4*, spine of ischium; *5*, rim of acetabulum; *6*, gluteus maximus muscle; *7*, labrum of acetabulum; *8*, trochanter major; *9*, iliofemoral ligament (Bertin); *10*, gluteus medius muscle; *11*, tensor fascia latae muscle; *12*, rectus femoris muscle and tendon; *13*, sartorius muscle; *14*, iliopsoas muscle; *15*, common femoral artery; *16*, common femoral vein; *17* pubis

acetabular fossa a small mass of fat can also be found. The normal, thin, low signal intensity trabeculae might become unusually prominent when increased forces are directed on the hip, along the lines of stress. On coronal images, the weight-bearing trabeculae course from the inferolateral to the superomedial aspect of the femoral head and may extend into the medial aspect of the neck. The signal intensity of bone marrow varies with age. Red marrow has a lower signal than yellow marrow. The conversion from hematopoietic to fatty marrow progresses from the distal to the proximal skeleton. In children and young adults the red marrow femoral neck has a lower signal intensity than the epiphyseal region. The femoral head and greater trochanter contain yellow marrow and have an intense signal at any age. The fatty marrow of the femoral neck and intertrochanteric region varies with aging; under 50 years of age, there is usually some red marrow in those regions. The distribution of the red and yellow marrow is generally bilaterally symmetrical and in some hematological disorders, there is marrow reconversion and hyperplasia.

The capsular structures insert on the intertrochanteric line, and the pubo-, ischio-, and iliofemoral ligaments are capsular thickenings (Figs. 4.1–4.3). The fibrocartilaginous acetabular labrum or cotyloid ligament can be identified as a small structure of low signal intensity, with a thick base of attachment and a thin and sharp free edge, best depicted in the case of joint effusion. The thin

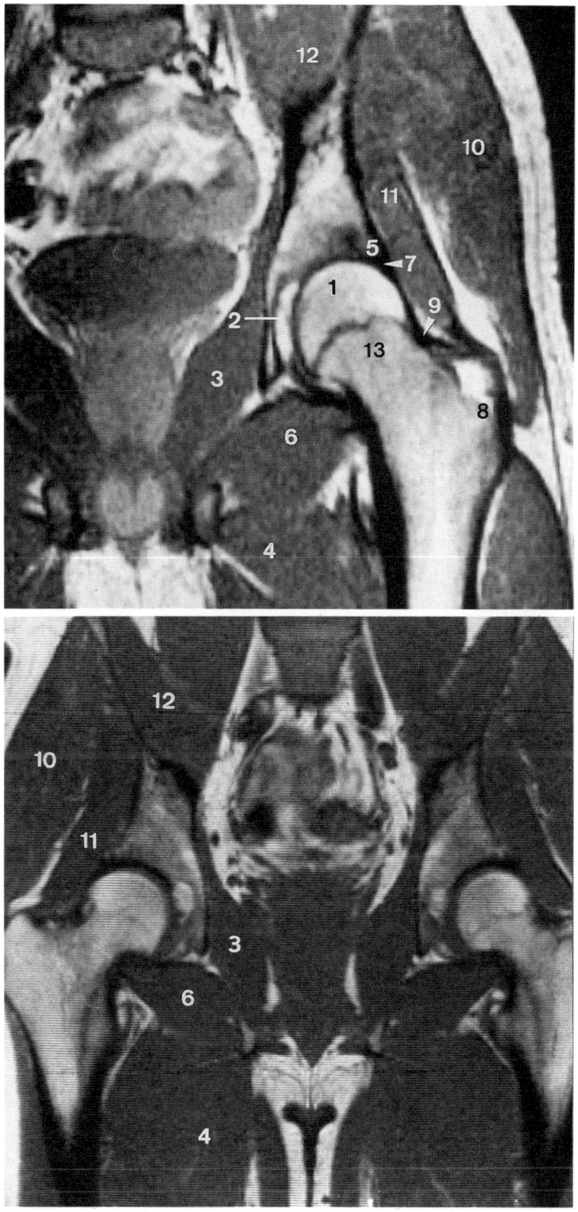

a

b

Fig. 4.2 a, b. Coronal sections. **a** A 15-year-old boy. **b** A 40-year-old man (bilateral view). *1*, head of femur; *2*, acetabular fossa; *3*, internal obturator muscle; *4*, adductor muscle group; *5*, rim of acetabulum; *6*, external obturator muscle; *7* labrum of acetabulum; *8*, trochanter major; *9*, iliofemoral ligament; *10*, gluteus medius muscle; *11*, gluteus minimus muscle; *12*, iliacus muscle; *13*, neck of femur

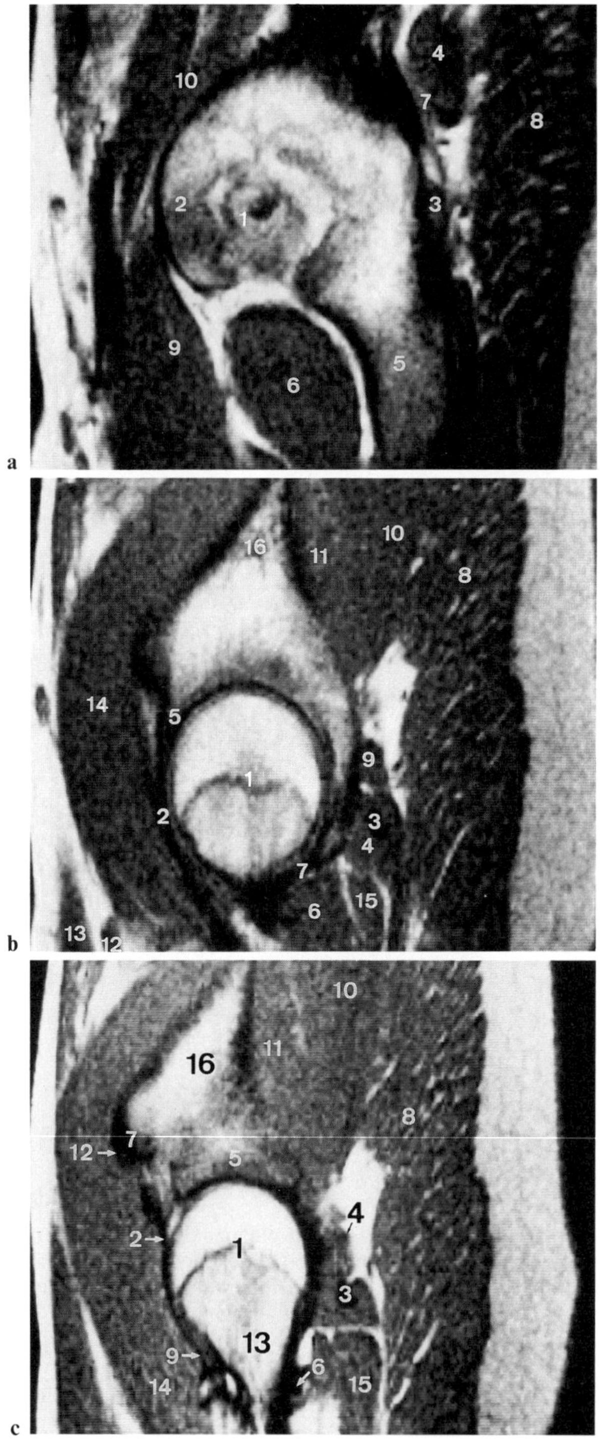
a
10
4
7
8
2
1
3
9
5
6
b
16
11
10
8
14
5
1
9
2
3
4
7
15
6
13
12
c
10
11
16
8
7
12
5
4
2
1
3
13
9
6
14
15

Fig. 4.3a–c

Fig. 4.3a–d. Sagittal sections of the hip joint. **a→d** Medial to lateral portion. *1*, Head of femur; *2*, capsular ligament; *3*, tendon of internal obturator muscle and gemelli muscle complex; *4*, tendon of piriformis muscle; *5*, acetabulum; *6*, tendon of external obturator muscle; *7*, anterior inferior iliac spine; *8*, gluteus maximus muscle; *9*, iliofemoral ligament; *10*, gluteus medius muscle; *11*, gluteus minimus muscle; *12*, tendon of rectus femoris muscle; *13*, neck of femur; *14*, iliopsoas muscle; *15*, quadratus femoris muscle; *16*, ilium

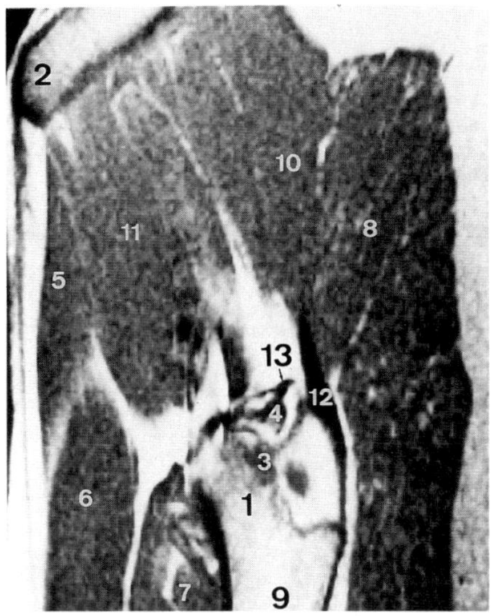

d

articular cartilage of intermediate signal intensity is not easy to evaluate on the MR images currently obtained. Three muscular groups are described in the hip. The anterior muscles comprise the iliopsoas muscle, sartorius muscle, and rectus femoris, which is a part of the quadriceps group. The posterior muscles include piriformis muscle, internal and external obturator muscles, superior and inferior gemelli muscles, and quadratus femoris muscle. The gluteus maximus, medius and minimus muscles, and tensor fascia latae muscle form the lateral muscle group. The medial muscle group contains pectineus muscle, gracilis muscle, and adductor muscles. Among the neurovascular structures the common femoral artery, bifurcating in superficial and deep arteries, the femoral vein, the femoral nerve, and the sciatic nerve are the most important.

The femoral head and the coxofemoral space are best evaluated on coronal and sagittal planes. Axial sections may provide additional information on the intra- and periarticular structures.

4.3 Avascular Necrosis of the Hip

Athletic trauma is one cause of avascular necrosis (AN), which can lead to severe hip deformity, dysfunction, and disability. A complete and detailed description of the pathogenesis of AN (Ehman 1988; Turner 1989; Mitchell 1989) is beyond the limit of this chapter, however it is important to mention that the diagnosis of AN at an early stage is very desirable because conservative and less invasive surgical measures are still possible. Early patient management and therapy may allow the infarct to heal while preventing undesirable

a 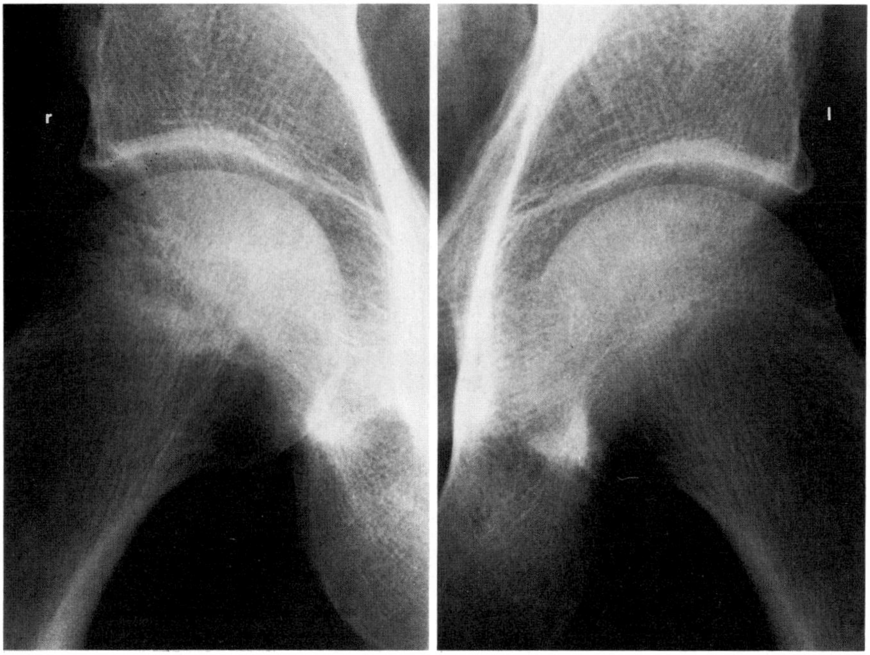 b

Fig. 4.4 a–e. Bilateral osteonecrosis in a 37-year-old nonprofessional soccer player with pain in the right hip. **a, b** Plain radiographs show patchy sclerosis of the right femoral head (*r*). Joint space is preserved. The left hip (*l*) appears normal. **c–e** The MR examination allows the detection of an early avascular necrosis of the femoral head in the contralateral hip which was not symptomatic at the time of examination. **c** Coronal T_1 image (SE 600/15) shows in the right an epiphyseal lesion; the signal is isointense with medullary fat surrounded by a band of low signal (*white arrow*). On the left hip, the developing lesion appears as a small intraepiphyseal crescent of low signal (*black arrow*). **d** On T_2-weighted image (SE 2100/70), areas of hyperintense signal are shown in the right femoral head (*white arrow*), outlined by the low-signal band. The important joint effusion has a bright signal intensity (*curved arrow*). The focal hypointense of the left femoral head remains unchanged. **e** Sagittal section of the right hip (SE 2100/22) shows posterior (*P*) extension of the lesion (*white arrow*) and persistent thickness of joint space (*arrowhead*). Joint fluid has an intermediate signal intensity (*curved arrow*)

complications such as secondary fracture, collapse, and destruction of the hip joint.

The staging system of AN of the femoral head, based on standard radiographic findings, has been described by Ficat and Arlet (1985):

- Stage I: Normal plain radiographs
- Stage II: Osteopenia, subchondral sclerosis, formation of subchondral cysts (occasionally)
- Stage III: Collapse of the femoral head
- Stage IV: Collapse of the femoral head, narrowed joint space, acetabular changes

Fig. 4.4c–e

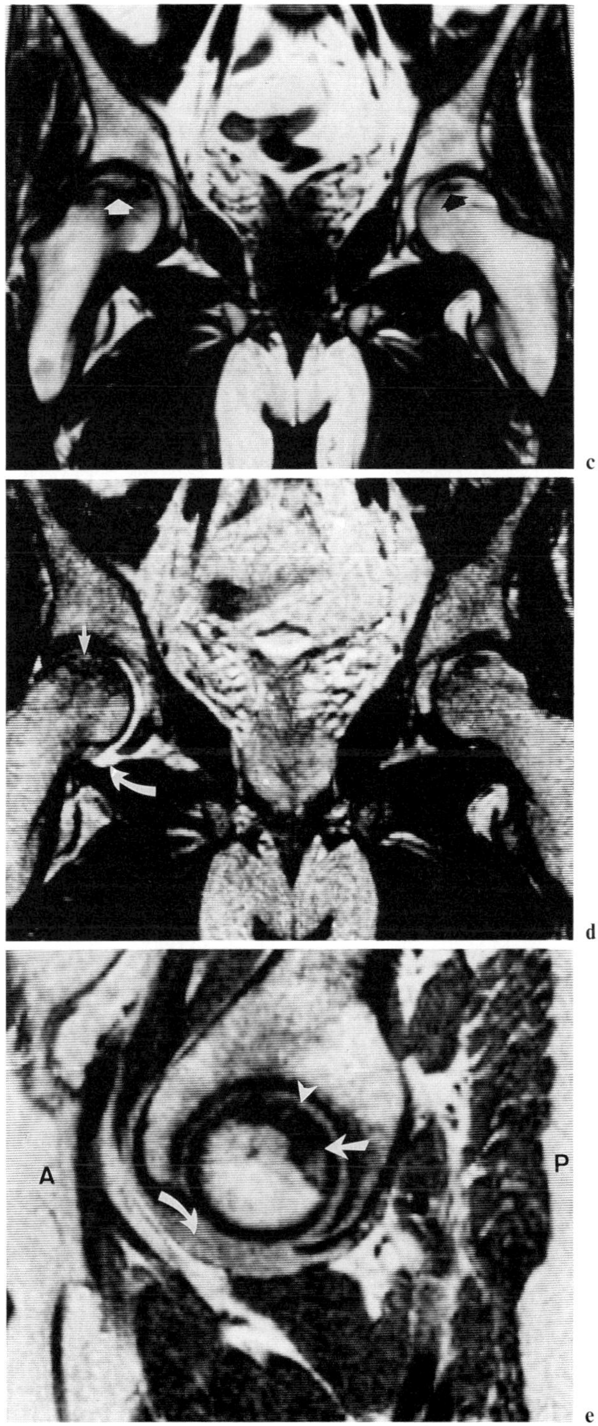

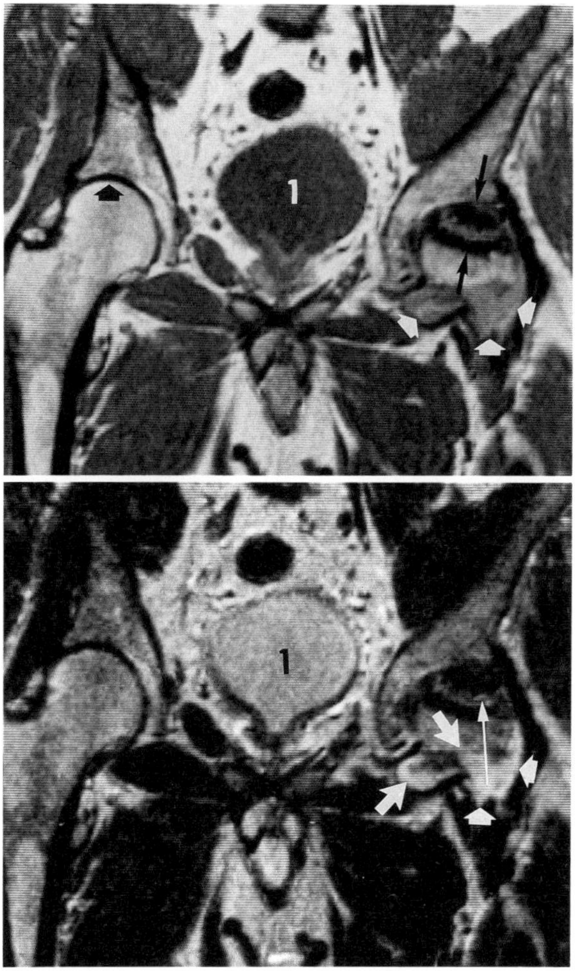

a

b

Fig. 4.5 a, b. Advanced stage of AN of the left hip in a 30-year-old rugby player. **a** On the coronal (SE 2100/22) image, the extensive lesion of the femoral head (*thin black arrows*) includes a superficial low-signal area surrounded by a zone of intermediate signal intensity. The necrotic lesion is well-demarcated from normal bone marrow by a low-signal band-like zone. **b** On the T_2-weighted (SE 2100/70) image, the dark signal of the superficial lesion remains unchanged and is consistent with a sequestrum. The signal intensity of the surrounding area is mildly brightening (*thin white arrow*); this area probably corresponds to a more active stage of the illness. Note: (a) The associated joint effusion of the left hip (*multiple white arrows*) with intermediate signal intensity on proton-density image (**a**) and high intensity on T_2 (**b**), appears rather inhomogeneous due to the presence of intraarticular "debris." (b) For comparison, the fluid in the bladder (*1*) is hypointense on proton density, hyperintense on T_2. (c) The right femoral head is round and well defined (*black arrow*) with normal signal intensity

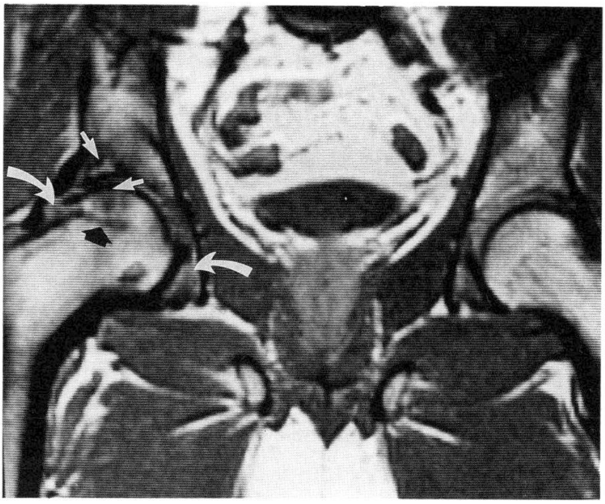

Fig. 4.6. Old fracture of the right femoral head in a young patient with persistent hip pain. Standard radiographs depict slight degenerative changes. MR examination is performed to exclude posttraumatic ischemic necrosis. The coronal T_1-weighted image (SE 2200/22) of the hips displays on the right a low-signal horizontal fracture line (*black arrow*). Subchondral cysts (*straight white arrows*) as well as sclerotic borders are depicted in both femoral head and acetabulum ("mirror" images). A banal intraosseous cystic lesion is also seen inferiorly. These early osteoarthrotic features are associated with an intraarticular effusion of intermediate signal intensity (*curved arrows*)

Recent studies (Turner 1989; Mitchell 1986; Lang 1988; Seiler 1989) demonstrate that MRI has a higher sensitivity in detecting changes of early AN than plain radiographs and is at least comparable to scintigraphy.

The MRI findings of early stages of AN are based primarily on the presence of characteristic focal abnormalities in the head of the femur (Fig. 4.4). These include small or large foci of decreased signal intensity on T_1-weighted images. The foci may be irregular or may appear as wedges, bands, lines, rings, or subchondral crescents (Turner 1989). Focal lesions are highly specific for the disease. Additional findings include diffuse signal abnormalities in the marrow of the neck and the intertrochanteric region of the femur. These present as hypointense on T_1- and hyperintense on T_2-weighted images. This is due to an increase in free water content attributed to the presence of edema or vascular congestion of the marrow. The late stage lesions of AN of the hip are well detected on plain films. They appear as areas of low signal intensity on T_1- and T_2-weighted MR images (Fig. 4.5). In advanced degenerative osteoarthrosis, MRI can help to differentiate the lesions from osteonecrosis. Degenerative changes of the hip joint, including subchondral cysts and sclerotic changes, are typically described in both femoral head and acetabulum, while necrotic lesions are confined to the femoral head (Fig. 4.6).

As technical considerations, we suggest that coronal views complemented by sagittal views are the best way to detect and confirm osteonecrotic lesions.

Additional axial sections may be useful to depict the precise topography of advanced AN.

4.4 Reflex Sympathetic Dystrophy Syndrome of the Hip

Posttraumatic reflex sympathetic dystrophy syndrome (RSDS; algodystrophy, Sudeck's atrophy) of the hip in adult athletes is not uncommon. In many popular sports today, the risk of soft-tissue injury to the hip joint is fairly high (e.g., soccer, football, athletism, jogging, etc.) Therefore, the possibility of RSDS should be kept in mind in the differential diagnosis of athletes with hip pain. RSDS is mediated via a neurovascular mechanism and develops rapidly. Pain and soft-tissue swelling are the usual clinical findings (Chapman 1984). The early manifestations of RSDS on standard radiographs are nonspecific and difficult to differentiate from other pathologies, such as the early stage of necrosis or transient osteoporosis. Soft-tissue swelling, locoregional demineralization, bone resorption, and small periarticular erosions are the most common radiographic findings (Resnick 1989; Chapman 1984).

At this stage, the high sensitivity of MRI permits the detection of lesions consistent with RSDS. The MR images clearly feature the diffuse aspect of the marrow-based changes characteristic of RSDS: a diffuse area of signal loss in the femoral head, neck and proximal diaphysis on T_1-weighted sections. The signal is typically brighter on T_2-weighted images (Fig. 4.7; Sarrat 1969). It is important to recall that algodystrophic changes usually appear broad and ill-defined while osteonecrotic lesions are rather more focal.

4.5 Microtraumatic Lesions of the Hip

The classic radiological triad – break in the cortex, radiolucent fracture line, displacement of bone fragment – is generally well displayed on standard X-rays as signs of gross posttraumatic lesions. However, lesions of lesser degree may occur and are not detected by conventional radiographs. Recent evidence has indicated that MRI is a very reliable method for detection and evaluation of occult cartilage and bone injuries (Mink 1989). The hip is a very vulnerable joint and may be the site of diverse types of injuries. An accurate and prompt evaluation of hip pain in adult athletes is important. Three types of occult (microtraumatic) bone lesions have been efficiently detected on MR images: (a) bone bruise, (b) stress fracture, and (c) osteochondral fracture, in which radiographic findings are usually equivocal or nonexistent (Yao 1988; Mink 1989).

In recent studies on traumatic patients (Mink 1989), MR images show bone bruises as geographic, nonlinear areas of low signal intensity on T_1-weighted images, usually involving the subcortical bone. An increase in signal intensity

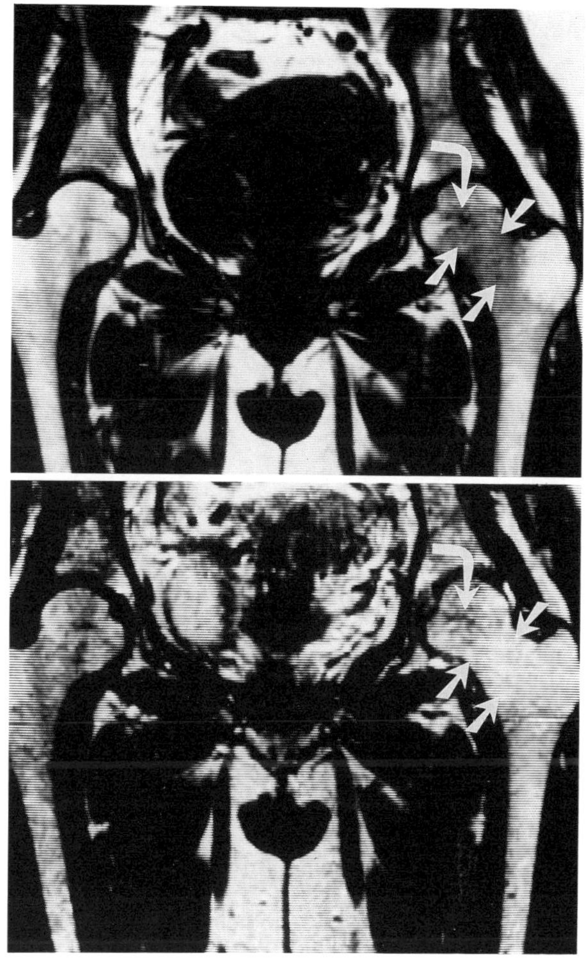

a

b

Fig. 4.7 a, b. Stress fracture with reflex sympathetic dystrophy syndrome. Young female athlete with severe unilateral hip pain. The standard radiographs were negative and the bone scan performed 3 weeks before showed a slight uptake in the left femoral head. **a** Coronal image (SE 600/15) depicts a focal zone of low signal intensity (*curved arrow*) across the left femoral head. A diffuse decrease of signal intensity (*straight arrows*), consistent with intrinsic medullary-based changes is seen in the femoral head and neck. **b** On T_2 (SE 2100/70), the signal intensity of this extensive area is typically increasing (*straight arrows*), while the fracture line remains low in signal (*curved arrow*). The diagnosis of stress fracture with secondary sympathetic dystrophy was strongly considered. Note that the contralateral femur shows a normal intense bone marrow on T_1 image (**a**), less intense on T_2 (**b**)

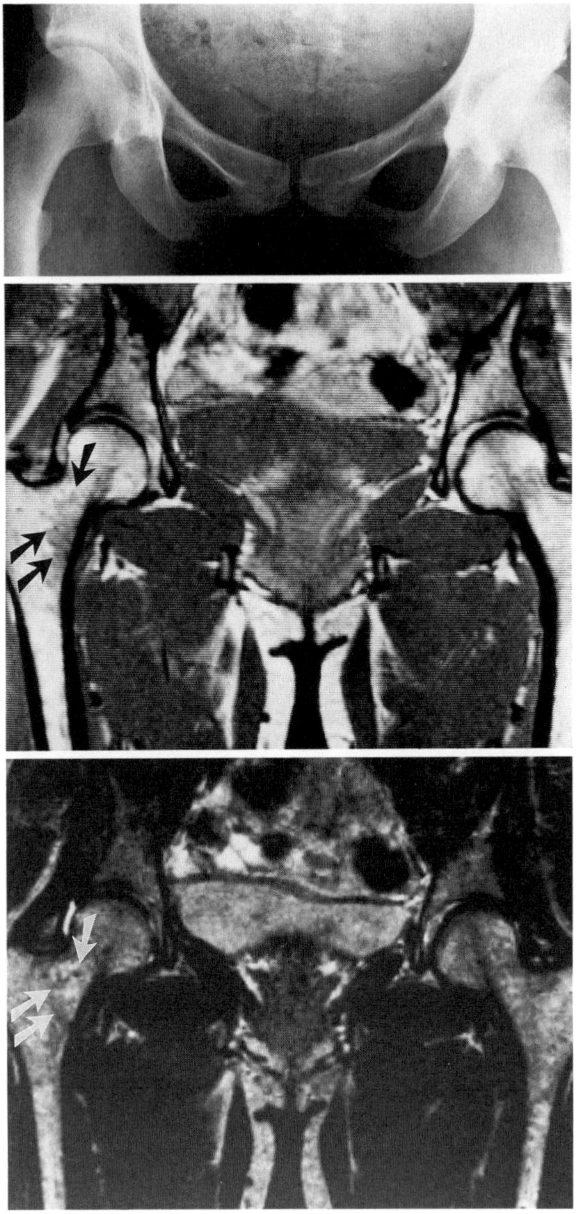

a

b

c

Fig. 4.8 a–c. Intramedullary bone contusion in the femur of a 28-year-old fervent female jogger with history of recent right hip pain. **a** The plain radiographs were negative at that time. **b** Coronal image (SE 2200/22) shows a geographic area of low signal intensity (*black arrows*) involving the subcortical bone at the inner part of the femoral neck. **c** An increase in signal intensity of the lesion is noted on T_2-weighted image (SE 2200/70), consistent with intraosseous edema (*white arrows*). Slight asymmetry of signal intensity must be carefully scrutinized

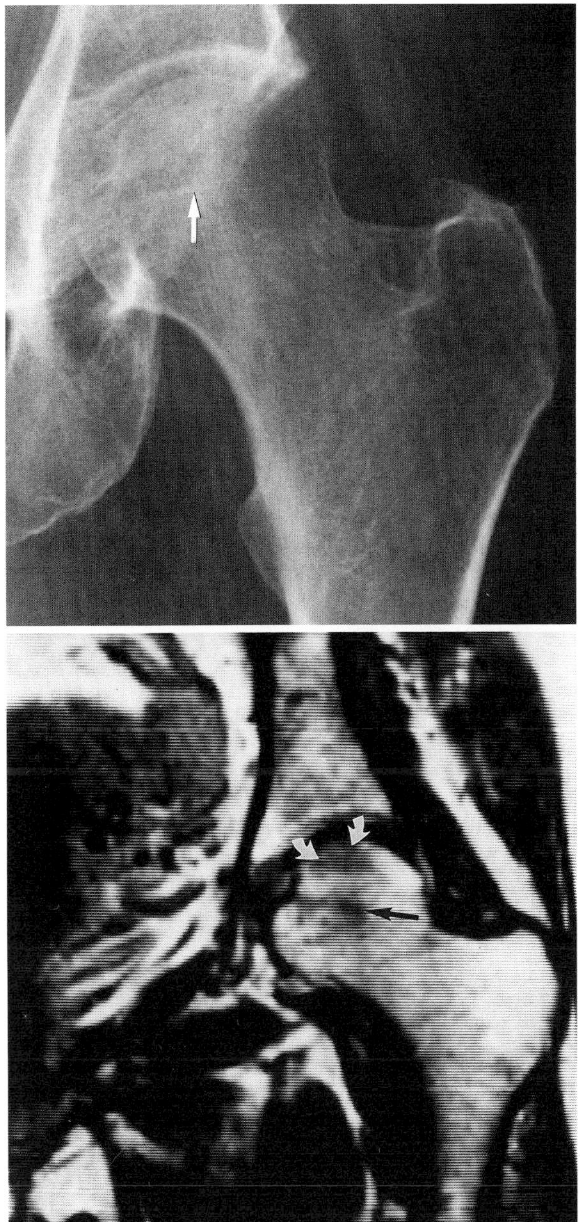

a

b

Fig. 4.9 a, b. Stress fracture of the left femoral head detected in a 35-year-old nonprofessional volleyball player with history of hip pain after a twisting injury. **a** On anterior or posterior X-ray view, slight heterogeneity of bone density is described in the right head with a possible area of focal sclerosis (*arrow*). **b** A MR coronal 3-mm-thick slice (SE 800/15) of the symptomatic hip shows a linear area of low signal intensity crossing the femoral head (*black arrow*). A subchondral hypointense zone is also depicted at the superior part of the femoral head (*white arrows*), which may correspond with "intraepiphyseal microimpactions"

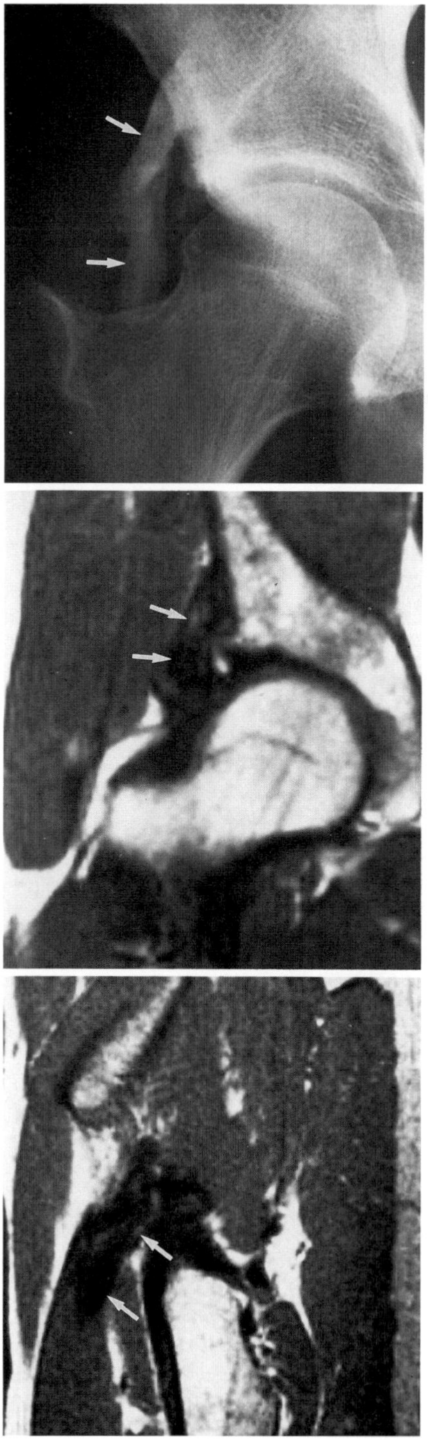

a

b

c

Fig. 4.10 a–c. Myositis ossificans in a 40-year-old professional referee who sustained several soccer injuries during the past 10 years. **a** Conventional radiograph shows a long linear calcified ossification at the lateral border of the coxofemoral joint line, consistent with myositis ossificans (*arrows*). MR coronal (**a**, 700/15) and sagittal (**b**, 2100/22) sections allows precise location of this area, which is lying at the proximal attachment of the rectus femoris muscle (*arrows*)

of the lesions is noted on T_2-weighted images (Fig. 4.8). Bone scintigraphy shows an increased focal activity in locations corresponding to areas of abnormality on MR images.

Stress fractures occur to bones which are repeatedly overloaded. Stress fractures of the femoral neck are seen especially in long-distance runners. The microfracture line is not visualized on conventional X-rays; this appears on MR images a linear, signal-loss area on T_1-weighted slices (Fig. 4.9). It is sometimes surrounded by a broader, ill-defined hypointense zone. The hypointense fracture line remains unchanged on T_2-weighted images. When reactive, focal edema exists, it appears hyperintense and adjacent to the fracture line (Mink 1989).

Osteochondral fracture means traumatic lesion affecting cartilage and adjacent subchondral bone, usually produced by a single trauma. This should not be confused with osteochondritis dissecans, which is caused by repeated traumatisms and additional vascular insufficiency, and where cartilage is initially normal. Occasionally the femoral head can be the subject of osteochondral fracture due to sport. The identification of chondral lesions on MR images is well documented (Ehman 1988; Turner 1989; Yao 1988; Mink 1989). The discrete zones of fracture are seen as hypointense zones on T_1- and T_2-weighted images. Coexistent medullary abnormalities – presumed edema – produce a hyperintense signal on T_2-weighted slices around the hypointense area.

Myositis ossificans is a soft-tissue mass, usually appearing after a hip contusion in young athletes who have active bone turnover. This condition is rare and is facilitated by strenuous activity and recurrent muscle tearing. The mass consist in fibrous tissue, cartilage, and bone. Calcific areas and heterotopic new bone formation are described at the periphery of the mass. The term myositis is not appropriate due to the absence of inflammation and muscular involvement in some cases. The lesion may cause muscular contracture and loss of hip motion. It can usually be identified by clinical examination and conventional radiography and must be differentiated from parosteal osteosarcoma (Resnick 1989; Fox in Nicholas 1986; Kulund 1988). MR images allow precise location and evaluation of the extent of myositis ossificans (Fig. 4.10). Associated muscle ruptures and hematomas can be detected, more accurately than ultrasound.

5 The Shoulder Joint

C. PIERRE-JEROME, M. SHAHABPOUR, and P. VAN ROY

5.1 Introduction

Most injuries to the shoulder joint occur in sports in which the athlete repeatedly throws or serves, for example, baseball or tennis. Other activities, such as swimming, skiing, and kayaking, which place a constant demand on the joint may also result in structural damage. Acute trauma or chronic repetitive movements are frequent causes of shoulder joint disorders as well. Cases of overuse injuries, stress fractures, subchondral bone damage, subluxation-dislocation have been reported in adolescent athletes, especially youngsters between 10 and 20 years of age who are in regular training for competitive sports. Athletes who are skeletally immature are more susceptible than adults to injuries resulting from microtraumata of growth cartilages. Lately, the incidence of sport-related shoulder injuries has increased due to higher levels of participation in several sports.

Adequate patient management and full rehabilitation requires efficient diagnostic techniques and accurate clinical examination.

5.2 Imaging of the Injured Shoulder

Diagnostic imaging techniques are extremely important in the clinical evaluation of the athlete with a painful shoulder. Several methods are available: conventional radiography, fluoroscopy, contrast arthrography, computed tomography, and others. Lately MRI has been used to image the shoulder joint. With the use of specialized surface coils, improved quality images are obtained, and exquisite depiction of the anatomic structures has become possible. Besides of being a noninvasive, nonradiation modality, MRI allows high contrast, multiplanar imaging of the shoulder.

5.3 MR Anatomy

The shoulder complex is a musculoskeletal structure providing multiaxial movements. Shoulder joint motion is basically performed at the glenohumeral joint, where only 25%–33% of the humeral head surface at once articulates with the glenoid fossa. The rather shallow glenoid fossa is slightly enhanced by the glenoid labrum, a rim of fibrous cartilage. T_1-weighted MR images depict very well the shoulder joint anatomy. The hyaline cartilage shows a moderate signal intensity; on the other hand, the fibrous cartilage of the glenoid labrum is characterized by the absence of MR signal. The glenohumeral joint shows the motion characteristics of a ball-and-socket joint type. The extremely high mobility of the shoulder is due partly to the contribution of the joints of the pectoral girdle (the acromioclavicular and sternoclavicular joints). Their mobility is possible only with a sliding motion of the shoulder blade against the thoracic cage. Therefore a physiologic joint may be considered between the scapula and the thoracic wall. The range of shoulder motion is also enhanced by associated movements in the articulations of the vertebral column.

The muscles connecting the scapula to the thoracic wall and vertebral column are responsible for scapular motion. They are clearly visualized on coronal oblique and sagittal MR pictures as structures with moderate signal intensity (Figs. 5.1–5.4). The muscles contrast well with the surrounding fat tissues of high signal intensity and the dark lines of adjacent cortical bone. Shoulder motion is also controlled by the muscles between thorax and humerus. The muscles and tendons of the rotator cuff (subscapularis, supraspinatus, infraspinatus, and teres minor) play a substantial role in the stability of the shoulder joint. Stability is provided primarily by the joint capsule, strengthened by capsular ligaments in front and at the upper pole of the joint. If not thoroughly unified with the joint capsule, shoulder ligaments can also be depicted as structures without MR signal.

Several synovial-lined bursal sacs – including the subdeltoid and subscapularis bursae – also contribute to the motion mechanisms of the shoulder joint complex. A physiologic joint relationship exists between the roof of the shoulder (the acromion and coracoacromial ligament) and the head of the humerus. On MR images of the subacromial space, the dark structure of the supraspinatus tendon can be clearly visualized (Figs. 5.1, 5.4). Abnormal relationship between the acromion, coracoacromial ligament, subacromial bursa, and supraspinatus tendon can be depicted (Huber 1986; Kieft 1986; Middleton 1987; Seeger 1987).

Shoulder pathology is often related to morphologic changes in one or more anatomical components, affecting the kinesiological mechanisms of the shoulder. MRI allows a good screening of the relationship of muscles, tendons, and bursae with respect to the articulating bone segments. Coronal oblique images through the belly of the supraspinatus muscle and axial and sagittal images may be helpful in the documentation of normal shoulder anatomy (Figs. 5.1, 5.3–5.5). Oblique images offer more standardized sections through the highly mobile shoulder complex (Kieft 1986; Seeger 1987).

One of the difficulties encountered in MRI of the shoulder joint is the loss of signal intensity, due to off-centered registration of MR signals. Therefore, improvements in the signal to noise ratio can be obtained by the use of appropriated surface coil technique (the images presented here are performed

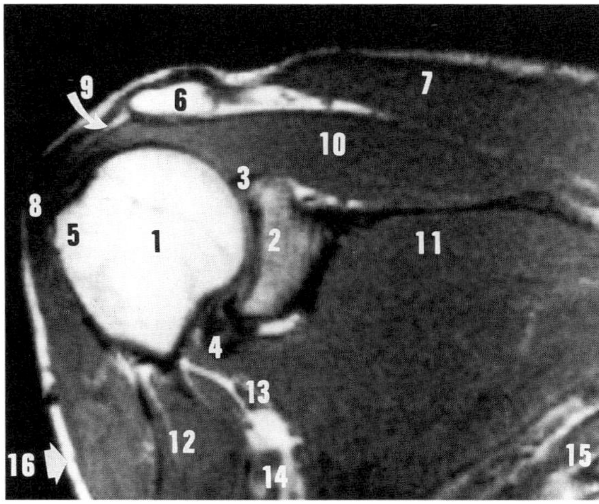

Fig. 5.1. Coronal oblique (SE 700/19, 3 mm thick) through the belly of the supraspinatus muscle. *1*, Humeral head; *2*, neck of the scapula with the glenoid fossa; *3*, glenoid labrum; *4*, axillary recess; *5*, greater tubercle; *6*, acromion; *7*, trapezius muscle; *8*, deltoid muscle; *9*, subdeltoid bursa; *10*, supraspinatus muscle; *11*, subscapularis muscle; *12*, triceps brachii muscle; *13*, posterior circumflex humeral artery; *14*, axillary artery and vein; *15*, intercostal muscles; *16*, subcutaneous fat tissue

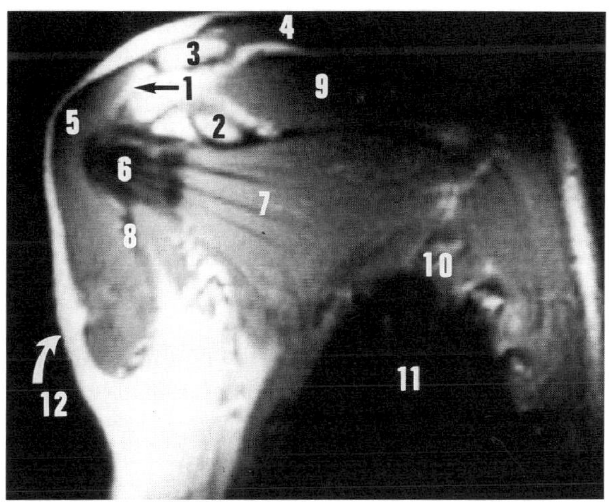

Fig. 5.2. Coronal oblique (SE 2100/26, 4 mm thick) through the subscapularis muscle. *1*, Acromion; *2*, coracoid process; *3*, clavicle; *4*, trapezius muscle; *5*, deltoid muscle; *6*, subscapularis tendon; *7*, subscapularis muscle; *8*, tendon of the long head of the biceps muscle; *9*, supraspinatus muscle; *10*, intercostal muscles; *11*, lung; *12*, subcutaneous fat tissue

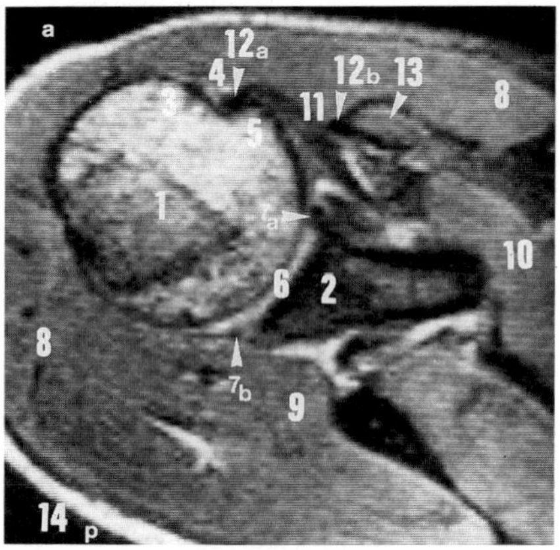

Fig. 5.3. Axial (3DFISP40 30/10, 1 mm thick) through the shoulder joint. *1*, Humeral head; *2*, neck of scapula; *3*, greater tubercle; *4*, bicipital groove; *5*, lesser tubercle; *6*, articular cartilage; *7a*, anterior glenoid labrum; *7b*, posterior glenoid labrum; *8*, deltoid muscle; *9*, infraspinatus muscle, *10*, subscapularis muscle; *11*, subscapularis tendon; *12a*, long head of biceps tendon; *12b*, short head of biceps tendon; *13*, coracobrachialis muscle; *14*, subcutaneous fat tissue; *a*, anterior; *p*, posterior

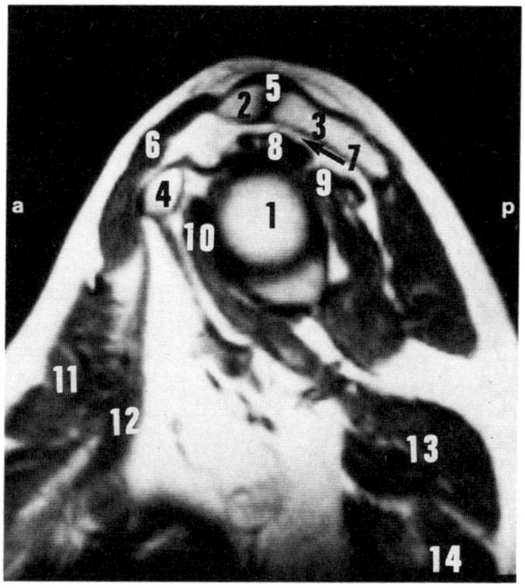

Fig. 5.4. Sagittal (SE 500/19, 4 mm thick) through the shoulder joint at the level of the humeral head. *1*, Humeral head; *2*, clavicle; *3*, acromion; *4*, coracoid process; *5*, acromioclavicular joint; *6*, deltoid muscle; *7*, subacromial space (with subacromial bursa); *8*, supraspinatus tendon; *9*, infraspinatus tendon and muscle; *10*, subscapularis tendon and muscle; *11*, pectoralis major muscle; *12*, pectoralis minor muscle; *13*, teres major muscle; *14*, latissimus dorsi muscle; *a*, anterior; *p*, posterior

using an experimental, mutually coupled Helmholtz resonator coil, designed by G. Annaert, V.U.B.; Kneeland 1986; Reiman 1988; Annaert 1989).

5.4 Common Sport-Related Injuries: MR Findings

Several types of lesions of the shoulder are seen in active adult and adolescent athletes. The aim of this chapter however, is to describe the MR findings in certain sport-related injuries of the soft tissues (tendons, ligaments, muscles) and intrinsic bony lesions that may be difficult to visualize on standard radiographs or by invasive techniques. These lesions include: rotator cuff tears, tendinitis (rotator cuff, bicipital), subacromial bursitis, shoulder impingment

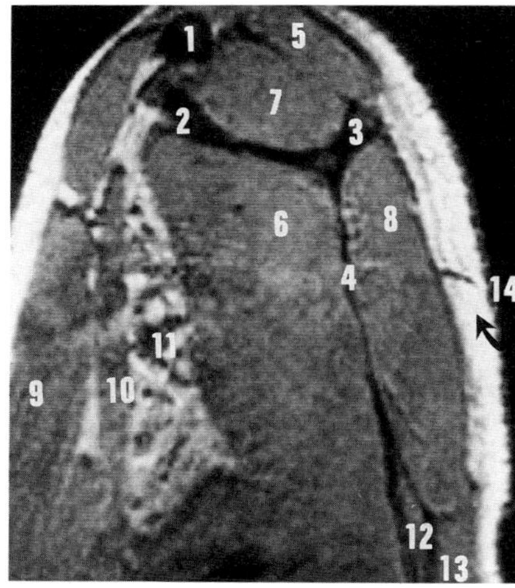

Fig. 5.5. Sagittal (3 DFISP40 30/10, 1 mm thick) through the lateral aspect of the scapular body. *1*, Clavicle; *2*, coracoid process; *3*, spine of the scapula; *4*, body of the scapula; *5*, trapezius muscle; *6*, subscapularis muscle; *7*, supraspinatus muscle; *8*, infraspinatus muscle; *9*, pectoralis major muscle; *10*, pectoralis minor muscle; *11*, neurovascular bundle; *12*, teres major muscle; *13*, latissimus dorsi muscle; *14*, subcutaneous fat tissue

syndrome, shoulder instability (anterior instability, inferior instability), and bone lesions (aseptic necrosis, microfractures, osteochondritis dissecans).

5.4.1 Rotator Cuff Tears

The rotator cuff of the shoulder consists of four muscles (supraspinatus, infraspinatus, subscapularis, and teres minor) that originate from the scapula and insert on the tuberosities of the humeral head. As they cross the glenohumeral joint, these muscles fuse and together form a musculotendinous hood that provides strength and dynamic stability to the joint (Seeger 1987; Kieft 1986; Middleton 1987).

Rotator cuff tears are both common and important. They may occur following acute trauma or repeated insults. Acute rotator cuff tears are commonly precipitated by lifting and by external rotation of the shoulder (a condition often seen in weight lifters and baseball pitchers). The tear may be partial or complete and associated with sudden pain and loss of motor power. Complete tear may accompany anterior dislocation of the glenohumeral joint. Most tears of the rotator cuff are overuse injuries. These injuries occur most often in athletes who repeatedly throw or serve, such as tennis players, pitchers, and volleyball players. In athletes over 35 years of age, tears may be induced by degenerative processes. Other sport activities that place a constant demand on the shoulder joint, such as swimming and skiing, also predispose to rotator cuff ruptures (Anderson 1988).

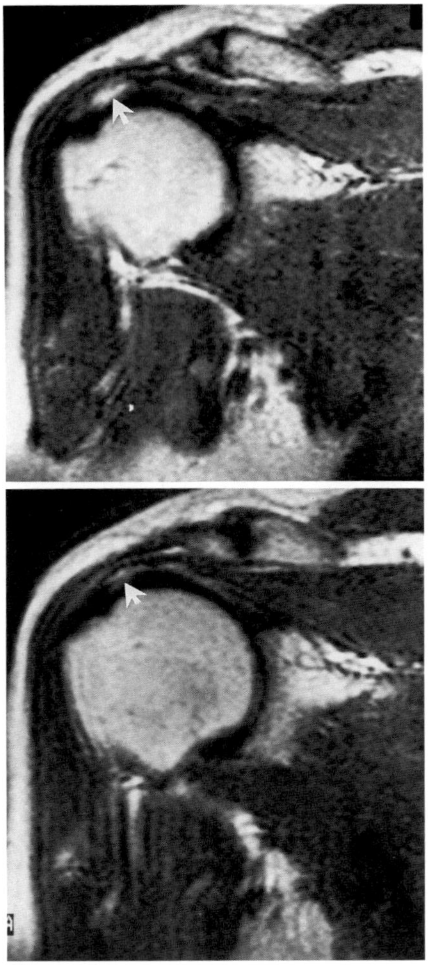

a

b

Fig. 5.6 a, b. Partial tear of supraspinatus tendon in a 39-year-old woman with shoulder pain. Oblique contiguous slices (SE 2200/22, 80) of the right shoulder display a small area of hypersignal in the midportion of the supraspinatus tendon (*arrowheads*), just above the lateral aspect of the humeral head. This lesion is consistent with a partial tear of the tendon and was also visualized by arthrography. (Courtesy of Drs. B. Van den Berg, J. Malghem, and B. Maldague; Cliniques Universitaires St Luc, Université Catholique de Louvain, Brussels)

Arthrography is the current modality for diagnosing rotator cuff tears (Sartoris 1988). However, this is an invasive technique, and partial tears may sometimes be difficult to visualize.

On MR images, the normal rotator cuff appears as a complex structure, with the central tendons of the rotator cuff muscles having low signal intensity and the intervening cuff having medium intensity on all pulse sequences (Kneeland 1987). Defects in the rotator cuff are filled with either fluid, granulomatous tissue, or hypertrophied synovium, giving a low or medium signal intensity on T_1-weighted images. An increased signal intensity of fluid or abnormal tissue at the site of the tear is seen on T_2-weighted images. Also, disappearance of the peribursal fat plane on MR images is considered by many authors (Mitchell et al. 1988) as a reliable sign of a rotator cuff tear.

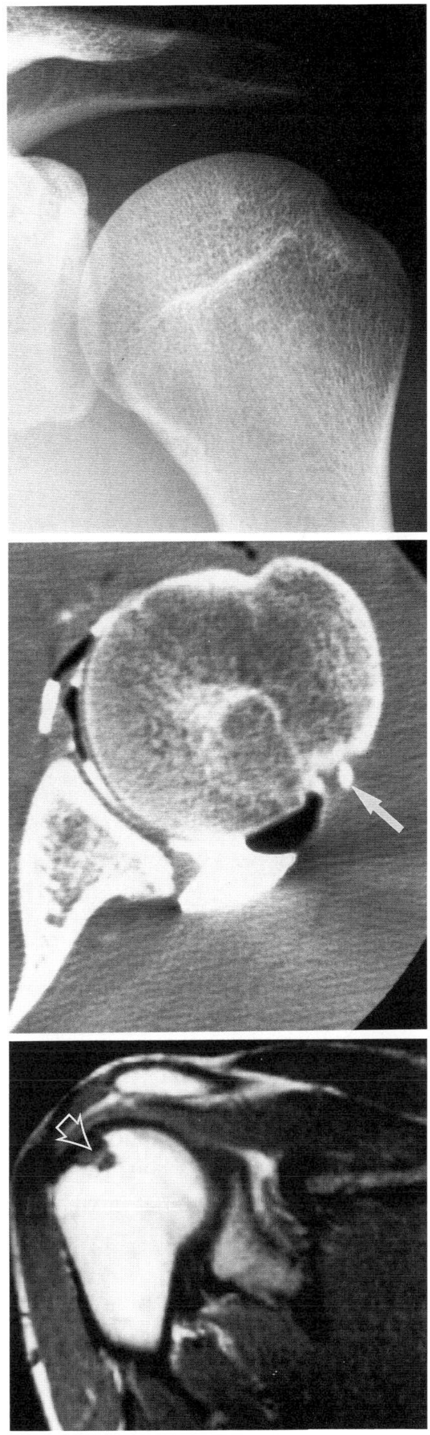

a b c

Fig. 5.7 a–c. Bone changes secondary to partial tear of the rotator cuff in a 28-year-old competitive tennis player, presenting chronic shoulder pain, particularly in abduction and rotation. **a** Plain radiograph appears normal. **b** Axial computed arthrotomography depicts slight contrast penetration within the insertion portion of the supraspinatus tendon (*arrow*), indicating the presence of a tendinous fissure. **c** Coronal oblique MR section (SE 700/19, 3 mm thick) displays a deep subcortical area with low signal intensity at the superior posterior portion of the greater tubercle, consistent with reactive intramedullary changes (*open arrow*)

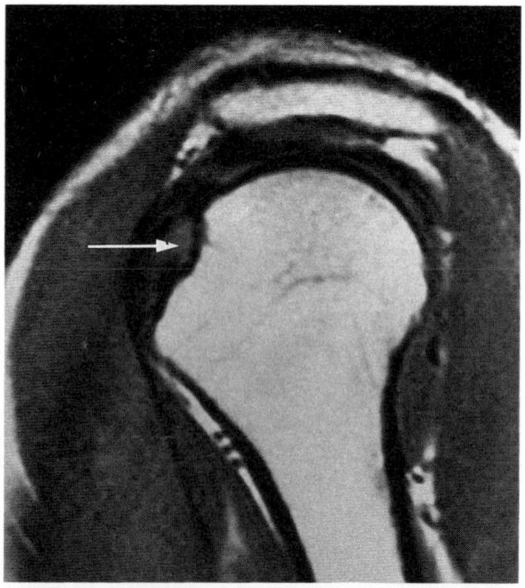

a

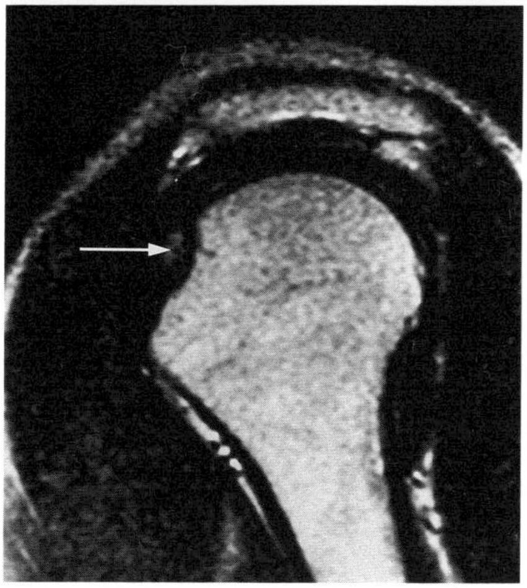

b

Fig. 5.8 a, b. Tendinitis of the subscapularis tendon in a 40-year-old man with persistent right shoulder pain. **a** Oblique slice (SE 2200/20) shows focal thickening of the subscapularis at the tendinous portion located close to the anatomic humeral neck, emiting an intermediate signal (*arrow*). **b** The signal intensity increasing on the T_2-weighted image (*arrow*), the lesion might correspond with tendinitis in active phase. (Courtesy of Drs. J. Malghem and B. Maldague; Cliniques Universitaires St Luc, Université Catholique de Louvain, Brussels)

Coronal and frontal oblique planes are useful in the detection of the rotator cuff tears, but additional sagittal sections may help to determine the extent of the lesion (Kneeland 1987). Small to moderate supraspinatus tears are usually not associated with retraction of the musculotendinous junction; however, in massive supraspinatus tear, the muscle may retract medially. (Figs. 5.6, 5.7).

5.4.2 Tendinitis

Tendinitis of the shoulder, in athletes, usually results from overuse. The most commonly encountered types are rotator cuff tendinitis and bicipital tendinitis.

In rotator cuff tendinitis the vascularity of the conjoined tendon of the rotator muscle group is such that there is an area of hypovascularity and limited circulation in the outer portion of the supraspinatus, which is further reduced with abduction and external rotation of the arm. This type of tendinitis occurs most frequently in such sports as tennis, volleyball, basketball, and handball that require repeated abduction and external rotation of the shoulder.

MR images reveal a medium-intensity signal within the substance of the tendon on T_1-weighted sections, which does not become hyperintense on T_2-weighting. On T_1-weighted images tendinitis cannot be differentiated from a small or partial tear of the supraspinatus. However, a well-defined focus of hyperintense signal within the tendon on T_2-weighted images is rather consistent with a tear (Fig. 5.8; Mitchell 1988, Seeger 1989b).

Bicipital tendinitis is rarely an isolated entity. It is usually associated with more widespread inflammation involving also the rotator cuff. The long biceps tendon may frequently subluxate during sport activities, precipitating recurrent bicipital tendinitis. The patient experiences a snapping or popping sensation as the tendon subluxates, with the shoulder in abduction, and internal or external rotation.

5.4.3 Subacromial Bursitis

The subacromial bursa can be involved in many pathologic processes, including rotator cuff tears. With complete tears of the rotator cuff, communication between the joint cavity and the subacromial bursa is established. Acute tears may result in hemorrhage or fluid accumulation within the bursa.

On MR images, the bursa generally shows high signal intensity on both T_1- and T_2-weighted images, probably due to high fat content of its synovium. In cases of subacromial bursitis there is a widening of the high intensity signal of the bursa medial to the region of depression and abrupt cut-off at the site of impingment (Seeger 1989). The bursal fluid gives a low to intermediate signal intensity on T_1-weighted sections and high signal intensity with T_2-weighting.

5.4.4 Shoulder Impingment Syndrome

Subacromial impingment is an important cause of shoulder dysfunction in athletes. This is a condition in which the supraspinatus tendon and subacromial bursa are chronically entrapped under the coracoacromial arc. As a result, the space for the bursa and tendon is reduced, and repetitive trauma to these structures – caused by vigorous overhead athletic endeavors – may lead to bursitis and rotator cuff injuries. The rotator cuff tendinitis may progress to fibrosis and rupture (Seeger 1988; Kieft 1988). Such injuries are produced by motions such as those involved in throwing a ball (baseball, basketball, tennis) or in javelin throwing. The diagnosis of shoulder impingment is based on clinical information and various imaging modalities: standard radiographs,

CT scan (Seeger 1988; Kieft 1988), ultrasound (Mack 1985), arthrography, fluoroscopy, and recently MRI (Seeger 1988; Kieft 1988; O'Neill 1989).

MRI offers the advantage of detailed imaging of the shoulder, and allows detection of several abnormalities associated with the shoulder impingement syndrome. Seeger et al. classify the MR findings of shoulder impingement syndrome into three different categories:

Type 1
- Subacromial bursitis manifested by bursal thickening
- Normal signal intensity of the supraspinatus tendon and muscle

Type 2
- High signal intensity of the supraspinatus tendon on T_2-weighted images, consistent with disruption of the cuff, and fluid infiltration within the cuff tear
- No muscle contraction

Type 3
- Abnormally high signal intensity of the supraspinatus tendon
- Retraction of the supraspinatus muscle, indicating complete rupture of the cuff.

When impingement syndrome is suspected, MRI examination of the shoulder can be performed in different planes, parallel or perpendicular to the glenoid surface. The use of surface coil is essential to obtain high-quality images. Frontal oblique scanning allows clear evaluation of the supraspinatus tendon while axial slices display well the glenohumeral joint.

5.4.5 Shoulder Instability

The glenohumeral joint is the most mobile joint in the body, and the trade-off for this exceptional mobility is an inherent instability. This is due mainly to the disproportionately small size of the glenoid cavity compared to the large humeral head. The two instability patterns seen most frequently in athletes are isolated anterior instability and inferior instability, or multidirectional instability. The problems associated with shoulder instability in the athlete include recurrent dislocation, pain, and weakness of the arm. Such a condition, if not properly treated, gradually worsens and leads finally to disability.

Inferior instability is extremely common in athletes who participate in sports in which overuse may occur, such as swimming. Also hard-working athletes who rapidly increase the intensity and duration of their workouts are more prone to develop instability (Anderson 1988).

As a noninvasive tool for evaluating the shoulder region, MRI has significant potential. MRI examination of the unstable shoulder, with the use of especially designed surface coils, provides detailed information about bone and soft tissue abnormalities. On MR images, labral lesions are seen as areas of increased signal intensity within the normal signal void of the fibrous labrum.

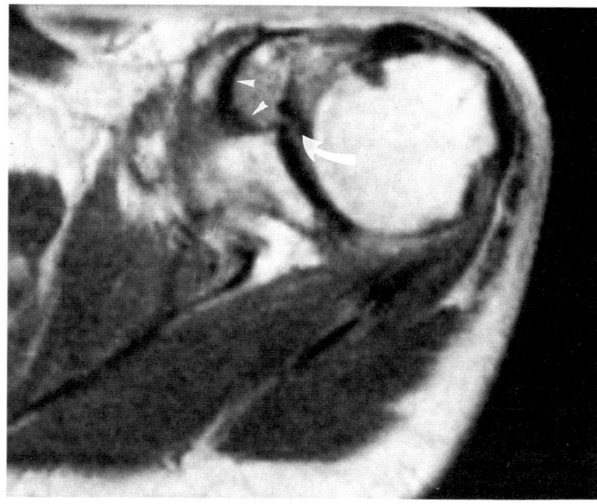

Fig. 5.9. Posttraumatic tear of the labrum in a 50-year-old tennis player, complaining of severe shoulder pain. Axial section (SE 2100/22, 5 mm thick) through the glenohumeral joint, anteriorly shows a large joint effusion distending the subcoracoid space (*arrowheads*). Signal intensity of the fluid is intermediate and appears heterogeneous. Small cartilaginous debris was found at arthroscopy. The anterior part of the glenoid labrum is also torn (*curved arrow*)

In cases of multidirectional instability, labral abnormalities are located anteriorly and posteriorly, on axial slices (Fig. 5.9; Seeger 1989). Tendons tears appear as areas of intermediate signal intensity on T_1-weighted images and become hyperintense on T_2-weighting. Frontal-oblique planes are well suited to depict Hill-Sachs lesions, which appear as focal depressions in the contour of the humeral head. When small and subtle, these depressed fractures are best seen on axial sections (Seeger 1989). As most humeral head dislocations are anteroinferior, it is important to keep in mind that Hill-Sachs defects are generally located on the high posterolateral aspect of the head, above or at the level of the coracoid process, on axial views. Care must be taken not to mistake the normal posterolateral flattening of the inferior aspect of the humeral head for a Hill-Sachs deformity (Seeger 1989; Kieft 1988). Association of medial retraction of the subscapularis musculotendinous junction is a reliable indicator of tendon rupture (Seeger 1989; Kieft 1988).

5.4.6 Bone Lesions

The gross bone lesions of the shoulder are usually of traumatic origin and are well displayed on standard radiographs or on CT scan. The role of MRI in this context is limited to detection of small intrinsic bone lesions not identifiable on conventional radiographs. Three types of lesions in particular deserve our attention: avascular necrosis, microfractures, and osteochondritis dissecans.

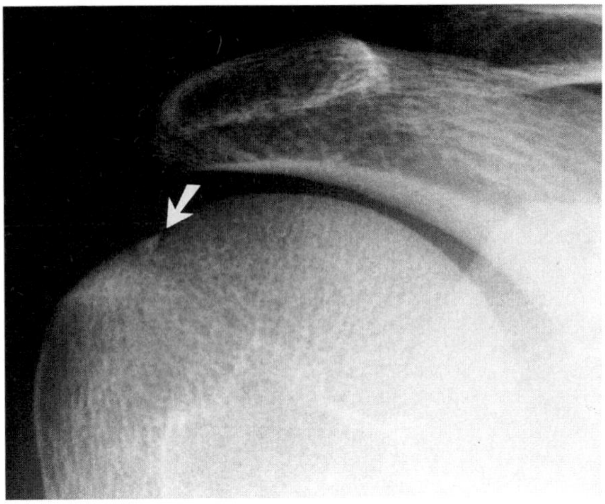

a

Fig. 5.10 a–d. Extensive medullary bone contusion of the humeral head. This 18-year-old patient presented with right shoulder pain starting 8 days after a fall from bicycle. **a** On a magnified plain radiograph, a very thin fissure associated with small focal break of the cortex (*arrow*) is seen at the junction of tuberculum majus and humeral head. Oblique sections (**b**, SE 450/20; **c**, SE 2200/20) show a large area of hypointense signal in the bone marrow of the tuberculum majus (*black arrows*). On the T_2-weighted image (**d**, SE 2200/80), the signal of the lesion becomes intense (*straight arrows*), consistent with intramedullary edema. A small intraarticular effusion (*curved arrow*) is visualized at the superomedial aspect of the humerus head. Focal increased signal intensity of the supraspinatus (*curved arrow*) is depicted at the junction of the tuberculum majus and the humeral head on proton-density image (**c**). This might be consistent with a partial rupture of the tendon. (Courtesy of Drs. J. Malghem, B. Maldague; Cliniques Universitaires St Luc, Université Catholique de Louvain, Brussels)

The humeral head is a common site for osteonecrosis. When the necrotic process is well established, the diagnosis can easily be made by standard radiographs. MRI is helpful in depicting the lesion at its early stages. The characteristics of the lesion on MR images are similar to those seen in other bones, such as the femoral head. These features are described elsewhere (Chap. 4).

Microfractures may be the cause of shoulder pain in active athletes. These may occur in sports that involve falls on the shoulder (e.g., football, soccer, ice hockey) or in more violent sport activities (e.g., judo, karate). A microfracture may coexist with other clearly visible injuries, such as the so-called shoulder separation, which is a sprain of the ligaments of the acromioclavicular joint. This injury is usually caused by falls directly on the point of the acromion, such as when a soccer player falls forward onto the anteriorly rotated shoulder. In cases of persistent shoulder pain, with negative standard radiographs and doubtful scintigraphy, MRI examination should be strongly considered to exclude a possible occult lesion. The high sensitivity of MRI to early, focal, marrow-based changes, makes it the modality of choice for the diagnosis of microfractures (Fig. 5.10). (See MRI characteristics of microfracture in Chaps. 4, 7).

Fig. 5.10 b–d

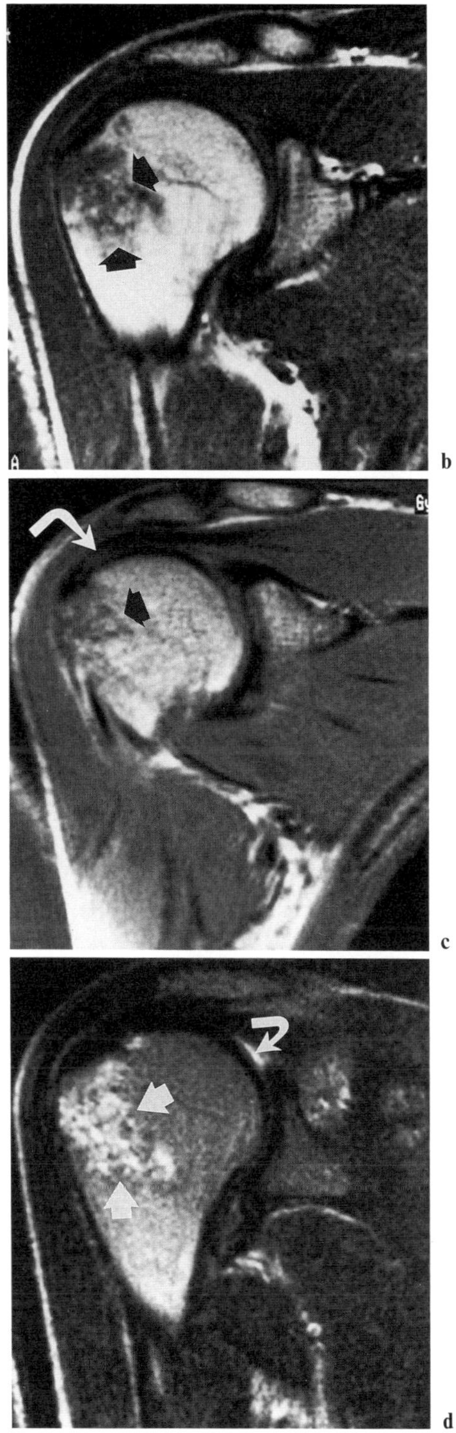

Osteochondritis dissecans is a rare cause of shoulder pain in athletes. Only few cases of osteochondritis of the shoulder have been reported in the literature, but it is found in some athletes, such as tennis players, as recently reported (Ishikawa 1989). It is important to remind the reader of the existence of such a rare lesion and the determining role of MRI in the diagnostic work-up of the painful shoulder in active professional athletes.

6 The Elbow Joint

C. Pierre-Jerome, M. Shahabpour, and K. Vandenborne

6.1 Introduction

The incidence of sports-related injuries of the elbow has increased over the past decade. Fortunately, advances have been made in understanding the mechanisms of injury and the biomechanics involved. Also, the use of more sophisticated techniques – such as MRI – has allowed a more accurate and prompt diagnosis of most sport-induced lesions of the elbow joint. Injuries to the elbow are seen in a variety of both contact and noncontact sports. In contact sports the elbow is frequently injured as a result of direct blows, falling on the outstretched or flexed arm. The consequent lesions are dislocations, isolated fractures, or other severe injuries. Injuries may also occur as an outcome of repeated stress of the arm, as in sports that involve throwing, hitting, pitching, bowling, or the use of a racquet. These athletes more often suffer damage to ligamentous and muscular insertions origins around the joint.

MRI of the elbow has a variety of potential roles: evaluation of suspected chondral lesions, investigation of posttraumatic bone marrow changes, and detection of abnormalities or compression of the neurovascular structures close to the joint.

6.2 MR Anatomy

Radiographic interpretation of the elbow requires a detailed recognition of all anatomical components of the joint. MRI enables evaluation of the complex elbow joint in coronal, sagittal, and axial planes (Mink 1989; Bunnell 1987; Figs. 6.1–6.4). It provides a clear depiction of muscles and tendons, bone marrow, articular cartilage, and neurovascular elements.

6.3 Sport-Induced Lesions

A well-synchronized radiological evaluation of the traumatized elbow starts with standard radiographs. Swelling of the periarticular soft tissue and a

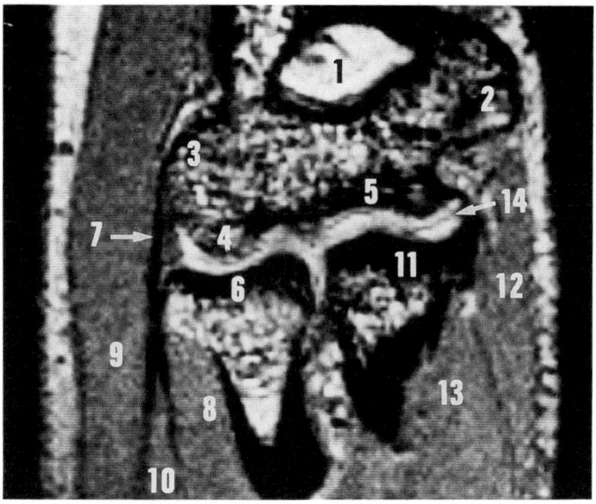

Fig. 6.1. Coronal 0.8-mm-thick section of the elbow joint (3DFISP40 30/10). *1*, Olecranon fossa; *2*, medial epicondyle; *3*, lateral epicondyle; *4*, capitellum; *5*, trochlea; *6*, radial head; *7*, lateral collateral ligament; *8*, supinator muscle; *9*, brachioradialis muscle; *10*, extensor carpi radialis longus and brevis muscle; *11*, coronoid process; *12*, pronator teres muscle; *13*, flexor digitorum profundus muscle; *14*, articular cartilage (two layers)

positive fat pad sign are strong indicators of existing traumatic lesions (Murphy 1977). We only describe here some of the frequent sport-related injuries of the elbow that are detectable by MRI: soft tissue injuries (lateral or extensor tendinitis, medial or flexor tendinitis), osteochondral fractures, and osteochondritis dissecans.

6.3.1 Soft Tissue Injuries: Extensor and Flexor Tendinitis

Extensor tendinitis, also called lateral epicondylitis (tennis elbow), is the most frequently reported lesion of the elbow in athletes (Collins 1985). Both lateral and medial tendinitis are typical overuse injuries. These conditions develop with repeated pronation and supination of the forearm. Tendinitis on the medial side of the elbow is common in sports involving overhead throws such as javelin and baseball, but it also occurs while serving a tennis ball or hitting a slapshot in hockey.

Lateral or extensor tendinitis is more often developed in the leading arm in the golf swing, in the tennis backhand and the discus thrower. It usually occurs in the 30 to 50 years of age and is due to inflammation of the aponeurosis overlying the extensor carpi radialis and the extensor communis.

The most likely mechanical etiology appears to be related to overuse and overload, associated with lack of appropriate preconditionning. There is local tenderness at the area of the lateral epicondyle where the common finger

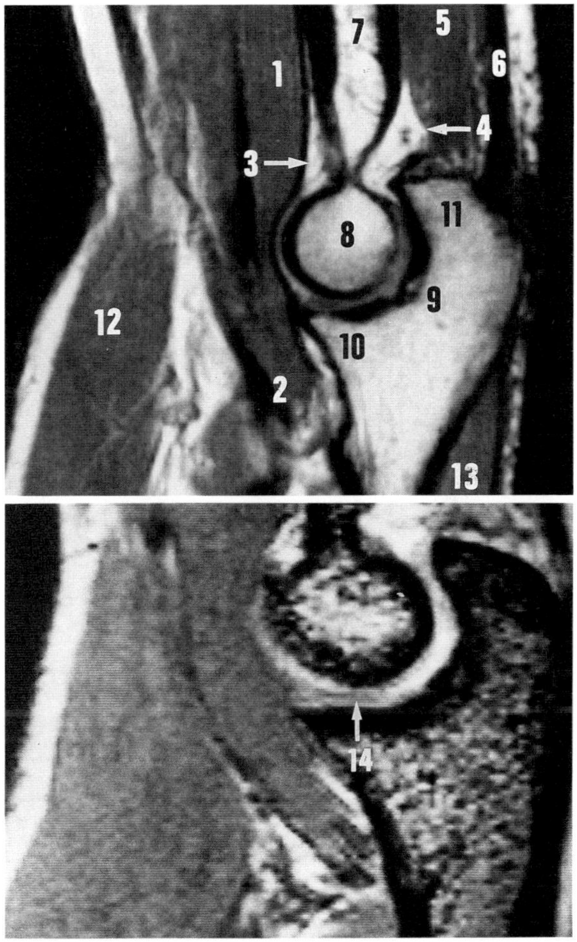

a

b

Fig. 6.2 a, b. Sagittal sections through the ulnar portion of the elbow joint. **a** SE 2200/22. **b** 3DFISP 40 30/10. *1*, Brachialis muscle; *2*, brachialis tendon; *3*, anterior fat pad; *4*, posterior fat pad; *5*, triceps muscle; *6*, triceps tendon; *7*, distal humerus; *8*, trochlea; *9*, trochlear notch; *10*, coronoid process; *11*, olecranon; *12*, pronator teres muscle; *13*, flexor digitorum profundus muscle; *14*, articular cartilage (two layers)

extensor muscles and extensor carpi radialis brevis muscle attach. Forced extension of the wrist against resistance reproduces elbow pain, which is lessened if the elbow is slightly flexed. The diagnosis of tennis elbow is mainly clinical; however, slight swelling of the soft tissues around the elbow can be seen on standard radiographs. Some atrophy of the extensor muscles of the forearm may occur in chronic cases.

On MR images, thickening of the tendon sheat manifests as a halo of hyperintense signal around the hypointense tendon on T_2-weighted sequences.

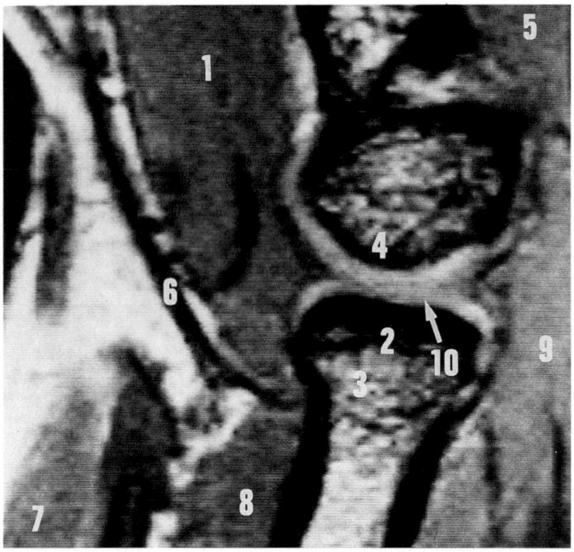

Fig. 6.3. Sagittal section through the radial portion of the elbow joint (3 DFISP 40, 30/10). *1*, Brachialis muscle; *2*, radial head; *3*, radial neck; *4*, capitellum; *5*, triceps muscle; *6*, biceps tendon *7*, brachioradialis muscle; *8*, supinator muscle; *9*, extensor tendons; *10*, articular cartilage (2 layers)

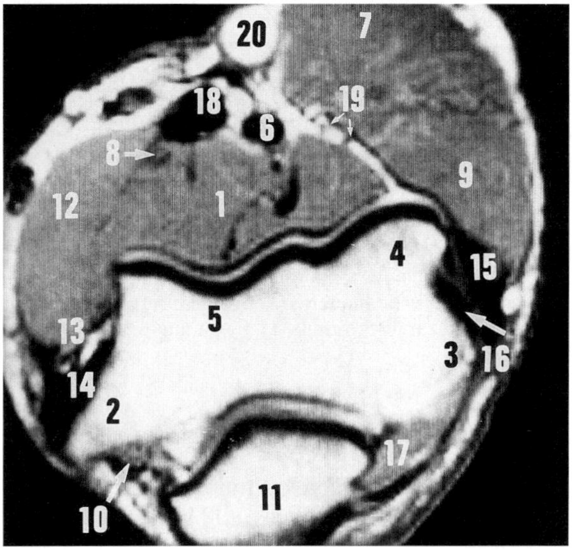

Fig. 6.4. Axial section of the elbow joint (SE 2200/22). *1*, Brachialis muscle; *2*, medial epicondyle; *3*, lateral epicondyle; *4*, capitellum; *5*, trochlea; *6*, biceps tendon; *7*, brachioradialis muscle; *8*, median nerve; *9*, extensor carpi radialis longus muscle; *10*, ulnar nerve; *11*, olecranon; *12*, pronator teres muscle; *13*, common flexor tendon; *14*, medial collateral ligament; *15*, common extensor tendon; *16*, lateral collateral ligament; *17*, anconeus muscle; *18*, brachial artery; *19*, superficial and deep radial nerve branches; *20*, cephalic vein

Fig. 6.5a–c. Osteochondral lesions of the right elbow in a 17-year-old man complaining of persistent pain. **a** Standard radiograph shows a subchondral bone defect. **b** Sagittal 3-mm-thick sections (SE 2400/22, 70) show a subchondral defect in the capitellum (*straight arrow*) associated with synovial hypertrophy (*arrowhead*) and joint effusion of heterogeneous high signal intensity on T_2 -weighted image (*curved arrows*). **c** Coronal 1-mm-thick section (3 DFISP40, 30/10) depicts associated chondral lesions of the capitellum as well as lesions of the cartilage covering the radial head (*thin arrows*). The lesions were confirmed by arthroscopy

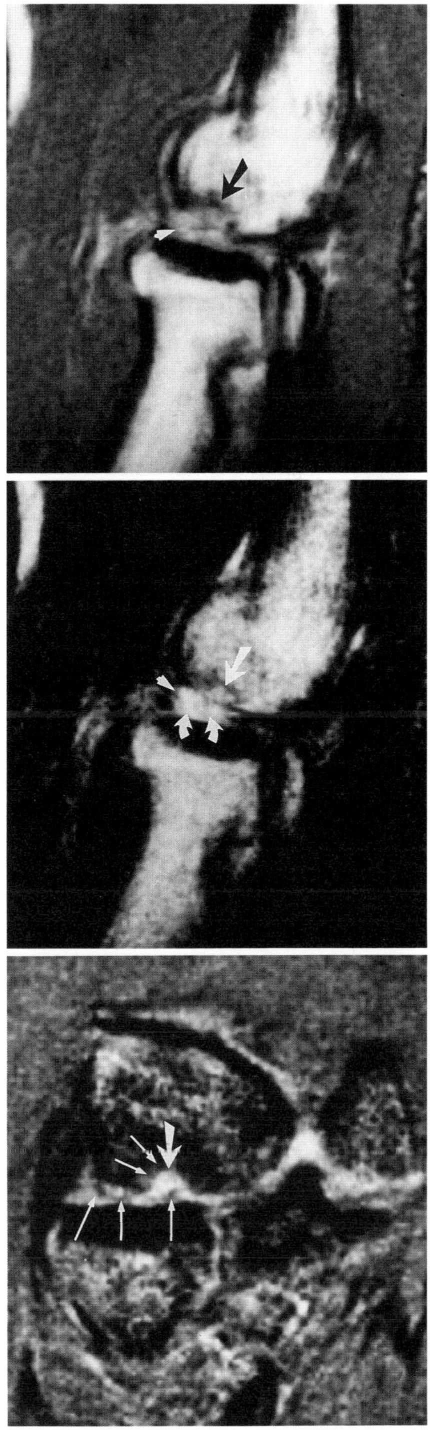

a

b

c

Periarticular and intraarticular fluid collections are also well visible on T_2-weighted images.

Flexor or medial tendinitis is much less common than extensor tendinitis. It involves the wrist and finger flexor tendons and the pronator teres tendon at their attachment to the medial epicondyle. The clinical presentation helps to establish the diagnosis. With radiology and MR, the signs are medially located and are similar to those described above.

6.3.2 Osteochondral Fractures

Compression injuries of the elbow may result in osteochondral fractures or impaction fractures. In the acceleration phase of throwing a baseball or javelin there is a large compressive stress upon the lateral side of the elbow between the radial head and the capitellum. These repetitive compressive forces may result in osteochondritis dissecans of the capitellum and osteochondral fractures. The final stage of elbow extension in the follow-through phase of throwing also causes degenerative changes at the tip of the olecranon. The same problem is also seen in weight lifting. These compressive lesions, usually not visible on plain radiographs, are frequent causes of painful elbow in professional baseball pitchers (Collins 1985). The player complains of pain in the elbow, occurring particularly at the end of throwing.

Coronal MR images are more useful in visualizing the cartilage and subchondral areas (Ehman 1988; Collins 1985). The osteochondral fracture is manifested by deformity of the cartilage and cortical bone. It appears as a poorly defined area of decreased signal intensity on T_1- and T_2-weighted images. A surrounding hyperintense zone of edema is detected on T_2-weighted images (Mink 1989; Yao 1988; Fig. 6.5).

6.3.3 Osteochondritis Dissecans

Trauma is one of the most frequent causes of osteochondritis dissecans of the elbow joint. Pain and limited range of motion with loss of full extension of the elbow are found in the clinical examination of athletes, especially throwers. The routine four radiographic views of the elbow – antero-posterior, lateral and both obliques – may lead to a definite diagnosis. However, in some cases, plain films are unrevealing, and further radiographic work-up becomes necessary.

Multiplanar MR examination provides a detailed evaluation of the elbow joint. Loosening and fragment displacement are directly visualized (Mesgarzadeh 1987). The role of MRI is then crucial in determining the mechanical stability of the osteochondral fragments. An osteochondritis dissecans fragment may be either stable or loose. When grossly loose, the overlying cartilage is disrupted, and the fragment is displaced, requiring surgical removal or internal fixation. Stable lesions are usually treated conservatively.

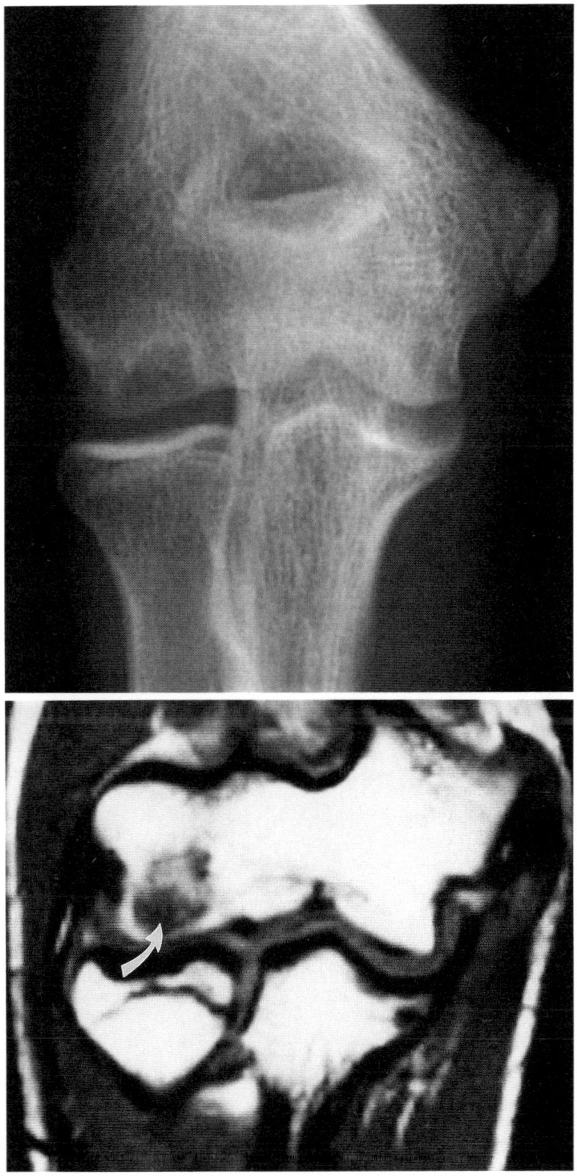

a

b

Fig. 6.6 a, b. Osteochondritis of the right elbow in a young professional swimmer with persistent pain. **a** Anteroposterior X-ray view shows a large subchondral bone defect in the capitellum, pointing out osteochondral pathology. **b** MR examination was performed before arthroscopy to evaluate global cartilage thickness and to exclude presence of intraarticular free fragments. Coronal image (SE 800/15) of the right elbow displays a well-defined lesion of low signal intensity (*curved arrow*) in the capitellum. Articular cartilage was thinned but no loose body was found on MRI and at surgery

On MR images, both stable and loose lesions appear as areas of low to moderate signal intensity on T_1-weighted images. Hypointense sclerotic changes at the margin of the parent bone are well delineated. The most obvious and reliable signs of loosening are displacement of the fragment and fluid collection at the interface of the fragment and the parent bone on T_2-weighted images (Teng 1984; Mesgarzadeh 1987; Fig. 6.6).

7 The Wrist Joint

C. Pierre-Jerome and M. Shahabpour

7.1 Introduction

The hand is a highly accident-prone area because of its frequent use in almost every sport. Furthermore, it is part of a protective reflex on impacts and in falls. Injuries to the wrist range from mild ligamentous injuries to dislocations that frequently require surgical reduction; major tendon injuries may terminate careers. Bone injuries of the wrist and hand are most frequently due to contact collisions or to falls on the outstretched hand. Consequently, it is not surprising that fractures and dislocations of the bones of wrist and hand are common in contact sports, such as rugby, soccer, wrestling, judo, and karate. However the same injuries also occur in events that require the swinging of an object, as in baseball, tennis, and golf, and in situations that require the release of an object with sudden wrist action, as in bowling, weight lifting, and certain track-and-field events. Since the presence of intrinsic bone lesions may not always be obvious on standard radiographs, the use of more sensitive, noninvasive techniques becomes imperative for accurate and definitive diagnosis.

MRI has proven to be efficient in the detection of occult cartilage and bone lesions (Mink 1989) and early posttraumatic medullary-based changes. Furthermore, lesions of fibrocartilages and injuries of fibrous structures such as tendons and ligaments are well displayed. Modifications – in aspect and location – of the median nerve can be easily detected. The wealth of diagnostic information provided by MRI, its ability to image in many planes, and the avoidance of ionizing radiation will made it the procedure of choice for the detection of many sport-related injuries of the wrist. Thorough knowledge of anatomy is essential for a logical approach of the traumatic wrist. More than any other technique, MR imaging exhibits the anatomical structures of the wrist, with the only exception of the small interosseous ligaments. These are occasionally – and partially – visualized due to their tiny size and geometrical orientation, better identified on thin 3 D images. MR examination of the wrists can be performed in axial, coronal, and sagittal planes (Figs. 7.1–7.4).

Traumatic lesions of the carpal joint occupy a large spectrum among common sport injuries that ranges from sprains to major fractures, dislocations, and fractures-dislocations. Description of most of these injuries is beyond the purpose of this chapter; we discuss here particularly the occult lesions, which are not visible on standard radiographs but are detected on MR images. Primary occult lesions include microfractures, medullary bone contu-

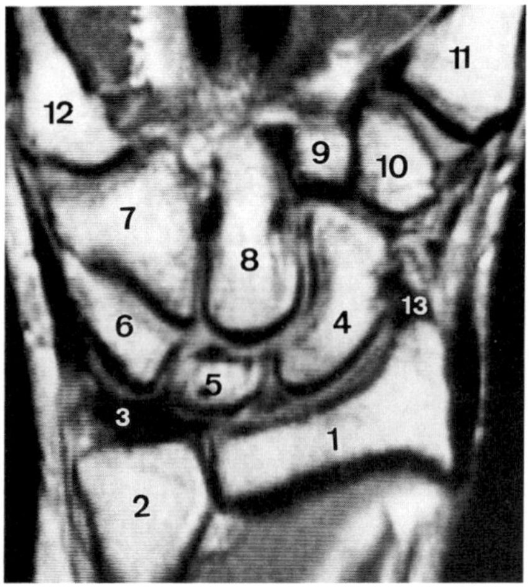

Fig. 7.1. Coronal section of the wrist (SE 2100/22). *1*, Radius, *2*, ulna; *3*, fibrocartilage ligament; *4*, scaphoid; *5*, lunate; *6*, triquetrum; *7*, hamate; *8*, capitate; *9*, trapezoid; *10*, trapezium; *11*, metacarpal 1; *12*, metacarpal 5; *13*, radioscaphoid ligament

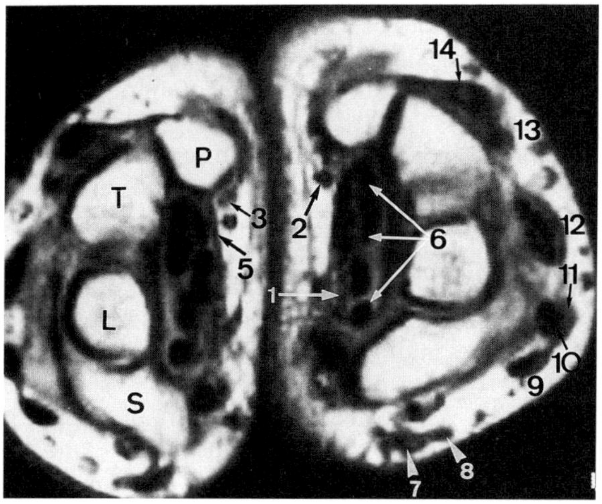

Fig. 7.2. Axial section of the wrists at the level of the proximal row (SE 2100/22). *P*, Pisiformis; *T*, triquetrum; *S*, scaphoid; *L*, lunate; *1*, median nerve; *2*, ulnar artery; *3*, ulnar nerve; *5*, flexor retinaculum; *6*, flexor tendons; *7*, abductor pollicis longus; *8*, extensor pollicis brevis; *9*, extensor carpi radialis longus; *10*, extensor carpi radialis brevis; *11*, extensor digitorum communis; *12*, extensor pollicis longus; *13*, extensor digiti minimi; *14*, extensor carpi ulnaris

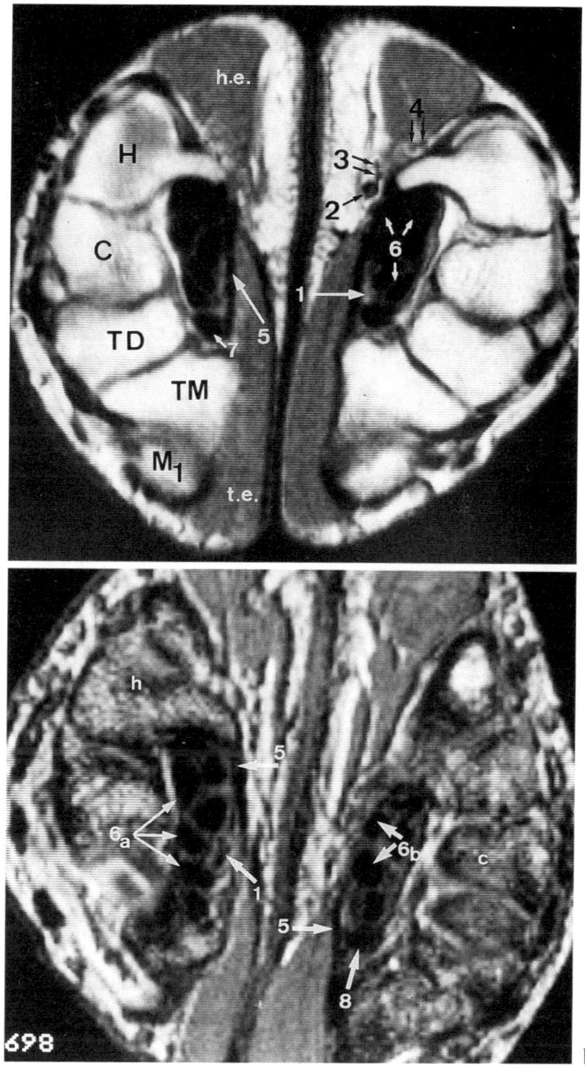

Fig. 7.3 a, b. Axial sections of the wrists at the level of the distal row. **a** SE 2100/22. **b** 3DFISP40 30/10. *t.e.*, Thenar eminence; *h.e.*, hypothenar eminence; *TM*, trapezium; *TD*, trapezoid; *C*, capitate; *H*, hamate; M_1, base of first metacarpal; *1*, median nerve; *2*, ulnar artery; *3*, superficial branches of ulnar nerve; *4*, deep branches of ulnar nerve; *5*, flexor retinaculum; *6*, flexor tendons; *7*, extensor tendons

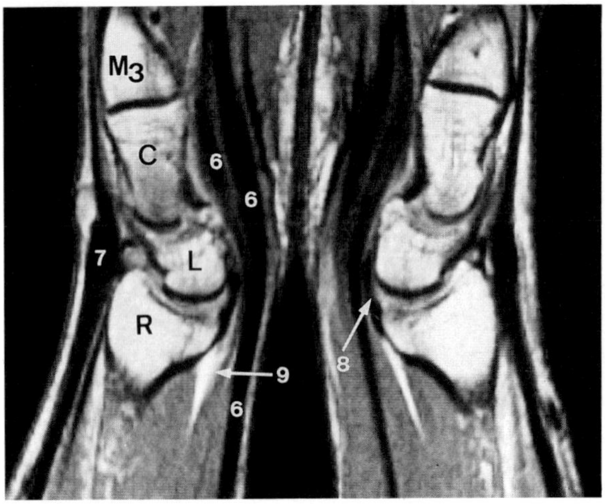

Fig. 7.4. Sagittal section of the wrists (SE 2100/22). *R*, Radius; *L*, lunate; M_3, third metacarpal; *C*, capitate; *6*, flexor tendons; *7*, extensor tendons; *8*, radiolunate ligament; *9*, pronator fat pad

sion, meniscal tears, and tendinitis and tenosynovitis. Among secondary lesions are necrosis (early and late stage), reflex sympathetic dystrophy, and nerve compression (including carpal tunnel syndrome).

7.2 Primary Traumatic Lesions

7.2.1 Microfractures or Occult Fractures

Injuries such as fractured scaphoid are very common and may be difficult to diagnose. The most common mechanism of injury to the scaphoid is a fall on the outstretched hand or an impact applied to the hand which forces the scaphoid against the radial styloid. A microfracture of the hook of the hamate can also be difficult to display on a radiograph. This fracture can result from a direct fall on the dorsiflexed hand, but it occurs more commonly from an injury sustained while swinging a tennis racquet, a baseball bat, or a golf club. Control is lost, and the butt of the handle strikes the hook of the hamate and fractures it. A linear lucency is the most frequently recognized sign of fracture on plain films. Fractures may also be identified by cortical discontinuity (a step-off) or increased density created by overriding fragments. Unfortunately, however, these signs are not always obvious. Microfractures generally remain undetected on plain radiographs.

These occult bone injuries can be well displayed by MRI. They may appear as (linear) areas of decreased signal on both T_1- and T_2-weighted images. They might be surrounded by a broader, poorly defined, and less dark area (of

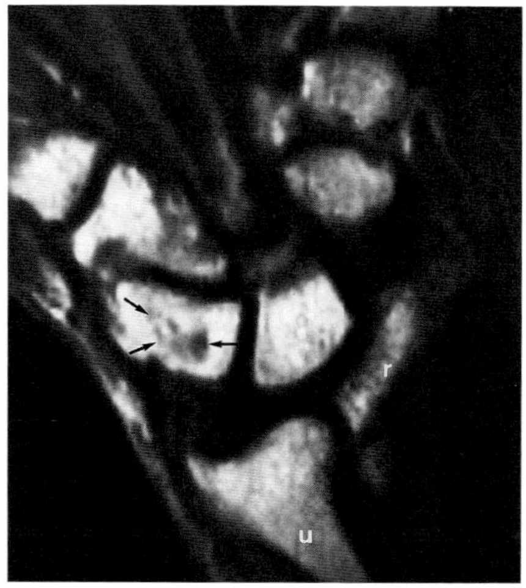

Fig. 7.5. Microfractures of the triquetrum in a 40-year-old female volleyball player with wrist pain. Conventional X-rays were negative. Coronal (SE 2100/22) 3-mm-thick section shows a hypointense medullary lesion at the triquetrum (*arrows*). *u*, Ulna; *r*, radius

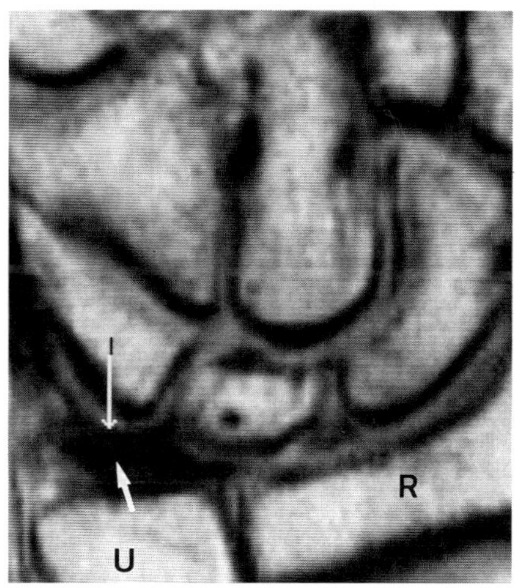

Fig. 7.6. Meniscal tear in a 28-year-old handball player with persistent wrist pain. Coronal section (SE 2100/22) of the right wrist shows a vertical tear (*arrows*) of the fibrocartilage ligament. *U*, Ulna; *R*, radius

presumed edema) on T_1-weighted images, which becomes brighter on T_2-weighted images (Fig. 7.5; Mink 1989, Yao 1988).

7.2.2 Medullary Bone Contusion

The contused medullary bone, or bone bruise, is not visible on conventional plain films. MRI has an unmatched capability to assess medullary-based changes. These are defined as areas of signal loss on T_1-weighted images, involving the subcortical bone, and appear hyperintense on T_2-weighted sections (Mink 1989).

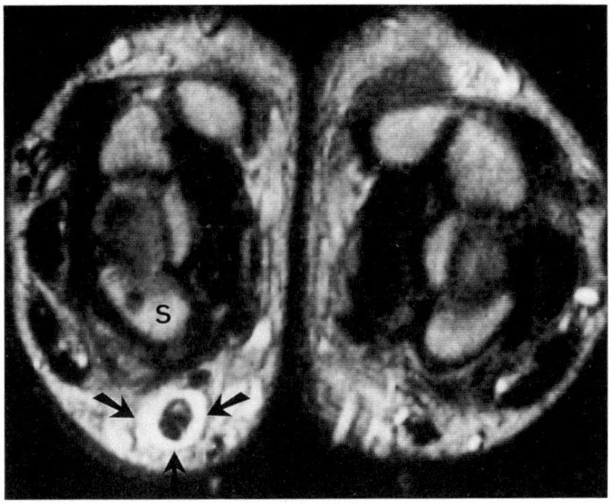

Fig. 7.7. Tenosynovitis in a 40-year-old professional typist with wrist pain and limited range of motion of the right thumb. Bilateral axial image (2100/70) of the wrists exhibits a high signal intensity halo (*arrows*) around the tendon of abductor pollicis longus. *S*, Scaphoid

7.2.3 Meniscal Tears

Mensical lesions are frequent causes of persistent painful wrist in athletes. Coronal MR images of the wrist are particularly useful in exhibiting meniscal tears. The carpal menisci, which are dark triangular-shaped structures on T_1 and T_2, appear disrupted. On T_2-weighted images, a variable amount of fluid collection can be seen, surrounding the damaged meniscus or crossing it (Fig. 7.6).

7.2.4 Tendinitis and Tenosynovitis

Tenosynovitis of flexor and extensor tendons is not infrequent in professional and nonprofessional athletes. It induces wrist pain and limitation of the range of motion of the carpus. On MR images, the affected tendon (flexor or extensor) appears surrounded by a halo of intermediate signal intensity on proton-density images, becoming hyperintense on T_2 (Figs. 7.7, 7.8). In cases

Fig. 7.8 a–c. Dorsal carpal cyst in a 28-year-old amateur tennis player who presented with ▶ right wrist pain and a moveable soft-tissue mass at the dorsum of his right carpus. **a** Bilateral sagittal section (2100/22) of the wrists shows an intermediate signal intensity mass (*C*), dorsal to the scaphoid. **b** On T_2-weighted sagittal image (2100/70), the lesion appears cystic, hyperintense, and multicameral. A small scaphoradial joint effusion is also visualized (*arrow*). **c** On T_2-weighted axial section, the lesion is located between the extensor tendons of the digitorum communis (*arrowhead*) and the extensor carpi radialis brevis (*curved arrow*). *S*, scaphoid; *R*, radius

Fig. 7.8 a–c

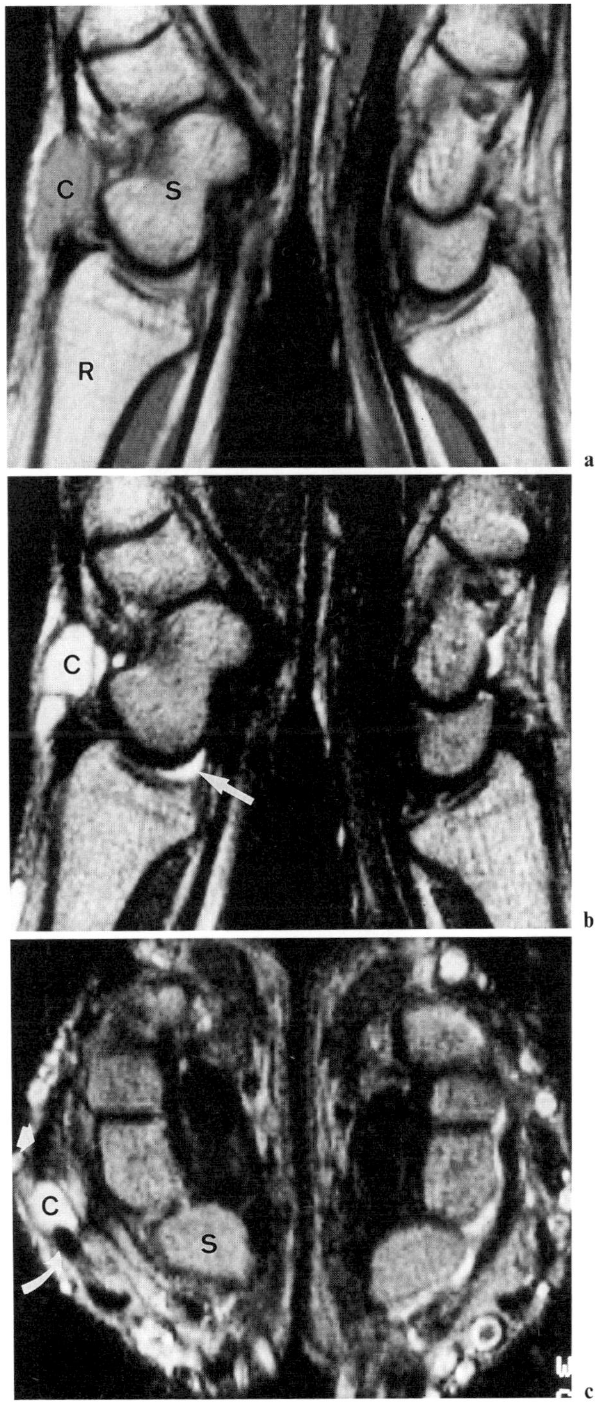

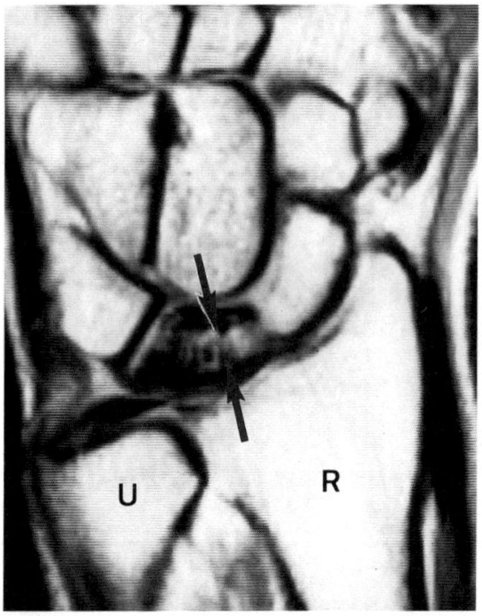

Fig. 7.9. Osteonecrosis of the lunate in a 35-year-old young female athlete with history of persistent wrist pain. Coronal (SE 600/15) 3-mm-thick slice of the right wrist shows a diffuse low signal intensity necrotic lesion of the lunate (*arrows*). *U*, Ulna; *R*, radius

of tenosynovitis of the flexor tendons – inside the carpal tunnel – the swelling of the tendon sheaths widens the gap between the tendons. Secondary focal irritation of the median nerve may produce carpal tunnel syndrome. Tennis athletism, bowling, and body building are among the sport activities often associated with tenosynovitis.

7.3 Secondary Lesions or Posttraumatic Complications

7.3.1 Osteonecrosis

Trauma-induced diminution or interruption of blood supply to a specific bone or bone segment results in ischemia and necrosis. The scaphoid is by far the most common carpal bone affected by traumatic osteonecrosis. Fractures of the waist or the proximal pole of the scaphoid are often complicated by avascular necrosis. Fractures of the distal pole usually do not result in necrosis because the blood supply remains adequate for rapid and reliable healing. The possibility of osteonecrosis in a posttraumatic, persistently painful wrist should always be kept in mind, even when plain films are negative. Indeed, symptoms are present before early bone marrow changes are evident on standard radiographs (Mitchell 1989).

Here again, MRI has shown its superiority over other diagnostic modalities. In cases of early necrosis, the affected area appears hypointense on T_1-weighted images, with a central zone of high signal intensity on T_2-weighted images (Mitchell 1989; Lang 1988). According to some authors, this hyperintense central zone is composed of a marrow segment that has not been reached by the repair process. In a more advanced stage, the necrotic zone appears rather hypointense on both T_1- and T_2-weighted images (Fig. 7.9). The late stage of necrosis is well demonstrated on standard radiographs.

7.3.2 Reflex Sympathetic Dystrophy

Sudeck's dystrophy is mediated via a neurovascular mechanism and is frequently associated with posttraumatic states. The wrist is commonly affected. The well-established dystrophic process has obvious radiological signs (demineralization, subperiosteal bone resorption, and small periarticular erosions). However, at an early stage the reflex sympathetic dystrophy may be symptomatic with no clear radiological evidences.

MRI can be helpful in the diagnosis of early-stage osteodystrophy. The demineralized zone has a characteristic low signal intensity on T_1-weighted sections and appears hyperintense on T_2-weighted images. The broad, diffuse aspect of the dystrophic lesions helps to differentiate them from osteonecrosis, which is more focal and better delineated (Sarrat 1988).

7.3.3 Carpal Tunnel Syndrome and Other Nerve Compressions

Nerve compression in the athlete's hand may be seen either in the median nerve in its tight carpal tunnel or in the ulnar nerve in Guyon's canal. The etiology of carpal tunnel syndrome and other nerve compressions is variable. Increase in intracanal pressure, focal median nerve compression, canal stenosis, and segmental swelling of the nerve are among the most frequently reported causes.

Compression of the median nerve may be due to trauma or repetitive powerful flexions, for example, in gymnasts. It has also been described in unexperienced rowers, body builders, and tennis and golf players. Because the lumbrical muscles may act as a space-occupying lesion within the carpal tunnel, an exaggerated development of these muscles may predispose occurrence of carpal tunnel syndrome (Middleton et al. 1987). Also, posttraumatic degenerative changes of the carpal bones, with resulting canal stenosis and focal inflammatory process within the canal, may induce carpal tunnel syndrome.

Bilateral, axial MRI examination of the wrists (Pierre-Jerome et al. 1989) clearly displays the carpal tunnel and its components, and modifications in the signal intensity, shape, and location of the median nerve are well appreciated. Pathological changes consistent with carpal tunnel syndrome and especially recurrent carpal tunnel syndrome – showing fat and fibrous proliferation, canal deformity, and swelling of the median nerve – are exquisitely demonstrated on axial T_1- and T_2-weighted images (Figs. 7.10, 7.11; Pierre-Jerome 1989; Middleton 1987).

Ulnar nerve compression may occur after fractures of the pisiform or the hamate. Direct trauma to the heel of the palm can cause scarring in Guyon's

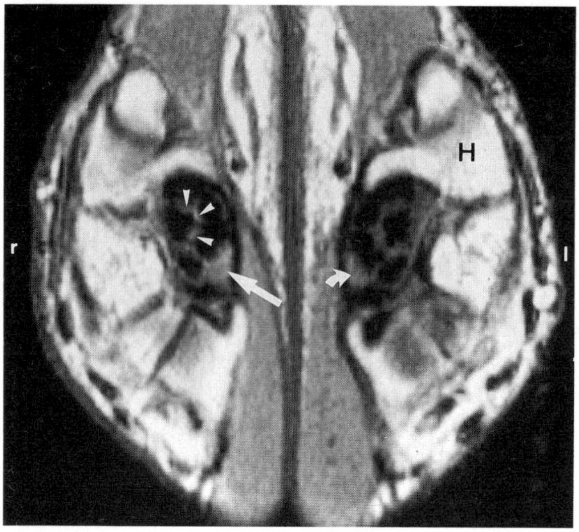

Fig. 7.10. Preoperative carpal tunnel syndrome in a 25-year-old active body builder. Bilateral axial image (SE 2100/22) displays on the right (*r*) a swollen, hyperintense median nerve (*straight arrow*) and abnormal bright signal intensity of the tendon sheaths (*arrowheads*). On the left (*l*), the median nerve (*curved arrow*) appears normal. *H*, Hamate

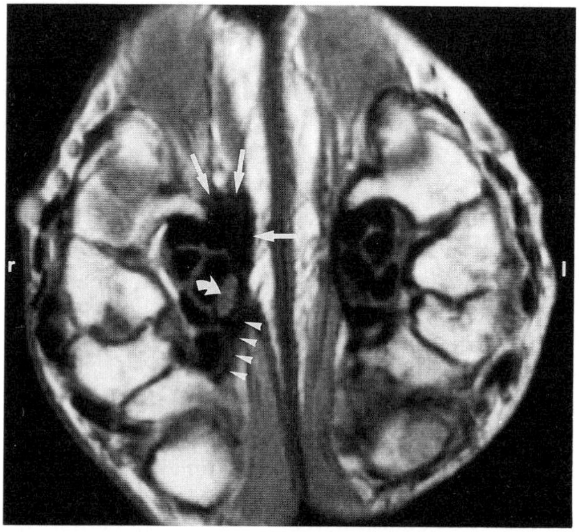

Fig. 7.11. Recurrent carpal tunnel syndrome. This axial image (SE 2100/22) illustrates the features of recurrent carpal tunnel syndrome in the right (*r*) wrist: hypointense fibrous tissue deposit (*arrows*), diffuse swelling of the median nerve (*curved arrow*), and incomplete resection of the retinaculum (*arrowheads*). On the left (*l*), normal median nerve and complete visualization of the flexor retinaculum

canal, which may lead to ulnar nerve compression. This condition is associated with sports such as judo, karate, gymnastics, and wrestling. Long-distance cyclists frequently develop numbness of the little finger and the ulnar half of the ring finger, with associated ulnar intrinsic muscle weakness. This is caused by their grip on the handlebars and constant pressure on the ulnar nerve in the Guyon's canal.

7.4 Conclusion

The wrist is possibly the most complicated joint in the human body and one of the most frequently injured. Because of the functional importance of the wrist, a prompt, accurate diagnosis after injury is therefore necessary. If undiagnosed, carpal lesions can lead to progressive limitations of motions, late degenerative intercarpal and radiocarpal arthritis, chronic pain, and disability. MRI proves to be the best technique for detecting occult posttraumatic lesions of the wrist, especially with three-dimensional techniques. The future of this technique, despite its high cost, is more than promising. Further development of new devices (for example, more refined surface coils and three-dimensional facilities) will allow an earlier diagnosis of occult carpal lesions.

References to Chapters 2–7

Anderson TE (1987) Rehabilitation of common shoulder injuries in athletes. J Musculoskel Med 5 (12):15–26

Andrews J (1983) Overuse syndromes of the lower extremity. Clin Sports Med 2:137–148

Baker DG, Schumacher HR Jr, Wolf GL (1985) NMR Evaluation of Synovial Fluid and Articular Tissues. J Rheumatol 12:1062–1065

Baker LL, Hajek PC, Bjorkengren A, Galbraith R, Sartoris D, Gelberman RH, Resnick D (1987) High-resolution MR imaging of the wirst: Normal anatomy. Skeletal Radiol 16:128–132

Bassett FM, Soucacos PN, Carr WA (1978) Jumper's knee: Patellar tendinitis and patellar tendon rupture. In: Funk FJ Jr (ed) Symposium on the athlete's knee. Hilton Head, CV Mosby, St Louis, pp 96–106

Bellon EM, Keith MW, Coleman PE, Shah ZR (1988) MR imaging of internal derangements of the knee. Radiographics 8:95–118

Beltran J, Noto AM, Herman LJ, Mosure JC, Burk JM, Christoforidis AJ (1986a) Joint Effusions: MR Imaging. Radiology 158:133–137

Beltran J, Noto AM, Mosure JC, Bools JC, Zuelzer WA, Christoforidis AJ (1986b) Meniscal Tears: MR Demonstration of Experimentally Produced Injuries. Radiology 158:691–693

Beltran J, Noto AM, Mosure JC, Weiss KL, Zuelzer WA, Christoforidis AJ (1986c) The Knee: Surface Coil MR Imaging at 1.5 T. Radiology 159:747–751

Beltran J, Noto AM, Mosure JC, Shaman OM, Weiss KL, Zuelzer WA (1986d) Ankle: Surface coil MR imaging at 1.5 T. Radiology 161:203–209

Beltran J, Noto AM, Herman LJ, Lubbers LM (1987) Tendons: High-Field-Strength Surface Coil MR Imaging. Radiology 162:735–740

Bentley G (1985) Articular Cartilage Changes in Chondromalacia Patellae. J Bone Joint Surg 67B (5):769–774

Berquist TH (1984) Magnetic Resonance Imaging, Preliminary Experience in Orthopaedic Radiology. Magn Reson Imaging 2:41–52

Berquist TH (1989) Radiology of the foot and ankle. Raven Press, New York, p 511

Berquist TH, Ehman RL, Richardson ML (1987) Magnetic Resonance of the Musculoskeletal System. Raven Press, New York, p 220

Bloem JL (1988) Transient Osteoporosis of the Hip: MR imaging. Radiology 167:753–755

Bodne D, Quinn SF, Murray L, Bolton T, Rudd S, Lewis K, Daines P, Bishop J, Cochran C (1988) MRI of chronic patellar tendinitis. Skeletal Radiol 17:24–28

Boland AL (1986) Soft tissue injuries of the knee. In: Nicholas JA, Hershman EB (eds) The lower extremity and spine in Sports Medicine. Mosby, St Louis, p 1085

Boven F, Bellemans MA, Geurts J, Potvliege R (1982) A comparative study of the patellofemoral joint on axial roentgenogram, axial arthrogram and computed tomography following arthrography. Skeletal Radiol 8:179–181

Boven F, De Boeck M, Potvliege R (1983) Synovial Plicae of the Knee on Computed Tomography. Radiology 147:805–809

Brostrom L (1965) Sprained ankles III: clinical observations in recent ligament ruptures. Acta Chir Scand 129:485–499

Bryan WJ, Wild JJ (1989) Isolated infraspinatus atrophy. Am J Sports Med 17:130–131

Bullough PG, Vosburgh F, Arnoczky SP, Levy IM (1984) The menisci of the knee. In: Insall JN (ed) Surgery of the Knee. Churchill Livingstone, New York, p 828

Bunnell D, Fisher DA, Bassett LW (1987) Elbow Joint: Normal anatomy on MR images. Radiology 165:527–531

Burk DL, Kanal E, Brunberg JA, Johnstone GF, Swensen HE, Wolf GL (1986) 1.5 T Surface-Coil MRI of the Knee. AJR 147:293–300

Burk DL, Dalinka MK, Kanal E, Schiebler ML, Cohen EK, Prorok RJ, Gefter WB, Kressel HY (1988) Meniscal and Ganglion Cysts of the Knee. AJR 150:331–336

Burke ER (1981) Ulnar neuropathy in bicyclists. Phys Sportsmed 9–53

Challis J (1986) Elbow and Forearm Injuries. In: Helal N et al. (eds) Sports Injuries and their Treatment. Chapman & Hall, London, p 499

Chapman S, Nakielny R (1984) Aids to Radiological Differential Diagnosis, Baillère Tindall. WB Saunders, London, p 375

Clemens M, Xhardez Y (1987) Le Genou Opéré, Prodim-Maloine, Bruxelles, p 178

Collins HR (1985) The Treatment of Shoulder and Elbow in Athlete Trauma Sports injuries: Mechanisms, prevention and treatment. Williams & Wilkins, Baltimore, p 877

Cotta H, Krahl H (1978) Das verletzte Kniegelenk des Sportlers. Dtsch Ärztebl 15:880

Cotty Ph, Proust F, Bertrand Ph (1988) Rupture de la coiffe des rotateurs. J Radiol 69 (11):633–638

Crues III JV, Mink J, Levy TL, Lotysch M, Stoller DW (1987) Meniscal Tears of the Knee: Accuracy of MR Imaging. Radiology 164:445–448

Daffner R (1987) Stress fractures. Skeletal Radiol 2:221–229

Daffner RH, Riemer BL, Lupetin AR, Dash N (1986) MRI in acute tendon ruptures. Skeletal Radiol 15:619–621

De Haven KE (1986) Injuries to the menisci of the knee. In: Nicholas JA, Hershman EB (eds) The lower extremity and spine in Sports Medicine. Mosby, St Louis, p 1085

Ehman RL, Berquist TH, Mc Leod RA (1988) MRI of the Musculoskeletal System: A 5-year appraisal. Radiology 166:313–320

Ficat RP (1985) Idiopathic Bone Necrosis of the Femoral Head. J Bone Joint Surg 67B:3–9

Ficat RP, Arlet J (1977) Ischémie et nécroses osseuses. Masson, Paris, p 223

Ficat RP, Philippe J, Hungerford DS (1979) Chondromalacia Patellae: A system of classification. Clin Orthop 144:55–62

Fleckenstein JL, Weatherall PT, Parkey RW (1989) Sports related muscle injuries: evaluation with MR imaging. Radiology 172:793–798

Fowler PJ, Messiem SS (1987) Isolated Posterior Cruciate Ligament Injuries in Athletes. Am J Sports Med 15(6):553–556

Frahm J, Haase A, Matthaei (1986) Rapid NMR Imaging of Dynamic Processes Using the FLASH Technique. Magn Reson Med 3:321–327

Frahm J, Hanicke W, Merboldt KD (1987) Transverse Coherence in Rapid FLASH NMR Imaging. J Magn Reson 72:307–314

Fullerton GD, Cameron IL, Ord VA (1985) Orientation of Tendons in the Magnetic Field and its Effect on T2 Relaxation Times. Radiology 155:433–435

Gagey N, Duvauferrier R, Frocrain L, Al Ghossein I, Ramee A (1987) Les bases physiques et anatomiques de l'interprétation d'une IRM de l'appareil locomoteur. Feuillets Radiologie 27(5):339–354

Gallimore GW, Harms SE (1986) Knee Injuries, High-Resolution MR Imaging. Radiology 160:457–461

Garrick JG, Requa RK (1988) The epidemiology of foot and ankle injuries in sports, Foot and Ankle Injuries. Clin Sports Med 7(1):29–36

Genety J, Brunet-Guedj E (1982) Le Genou du Sportif. Vigot, Paris, p 164

Gilula L, Oloff L, Caputi R, Destouet J, Jacobs A, Solomon M (1984) Ankle tenography: a key to unexplained symptomatology. Part II: diagnosis of chronic tendon disabilities. Radiology 151:581–587

Gylys-Morin VM, Hajek PC, Sartoris DJ, Resnick D (1987) Articular Cartilage Defects: detectability in cadaver knees with MR. AJR 148:1153–1157

Haase A, Frahm J, Matthaei D, Hanicke W, Merboldt KD (1986) FLASH Imaging, Rapid NMR Imaging Using Low Flip-Angle Pulses. J Magn Reson 67:258–266

Haggar AM, Froelich JW, Hearshen DO, Sadasivan K (1988) Meniscal abnormalities of the knee: 3D FT fast scan GRASS MR imaging. AJR 150:1341–1344

Hajek PC, Baker LL, Bjorkengren A, Sartoris DJ, Neumann CH, Resnick D (1986) High-resolution MR imaging of the ankle: normal anatomy. Skeletal Radiol 15:536–540

Hajek PC, Baker LL, Sartoris DJ, Neumann CH, Resnick D (1987) MR Arthrography: Anatomic-Pathologic Investigation. Radiology 163:141–147

Hammer M, Mielke H, Wagener P, Schwarzrock R, Giebel G (1986) Sonography and NMR Imaging in Rheumatoid Gonarthritis. Scand J Rheumatol 15:157–164

Handelberg F, Shahabpour M, Van Betten F, Pierre Jerome C, Casteleyn PP, Opdecam P (1989) CT arthrography and MRI of the patella. Acta Orthop Belg 55:331–338

Harms SE, Muschler G (1986) Three-Dimensional MR Imaging of the Knee using Surface Coils. J Comp Assist Tomography 10:773–777

Hartzman S, Reicher MA, Bassett LW, Duckwiler GR, Mandelbaum B, Gold RH (1987) MR Imaging of the Knee, Part II, Chronic Disorders. Radiology 162:553–557

Helal B, King JB, Grange WJ (1986) Sports injuries and their treatment. Chapman & Hall, London, p 499

Henry JH (1986) The patello-femoral joint. In: Nicholas JA, Hershman EB (eds) The lower extremity and spine in Sports Medicine. Mosby, St Louis, p 1085

Herman LJ, Beltran J (1988) Pitfalls in MR imaging of the knee. Radiology 167:775–781

Huber DJ, Sauter R, Mueller E, Requardt H, Weber H (1986) MR imaging of the normal shoulder. Radiology 158:405–408

Hughston JC, Barrett GR (1983) Acute anteromedial rotatory instability: long-term results of surgical repair. J Bone Joint Surg 65A:145–147

Hume E, McKeagh D (1984) Soft tissue ankle injuries: the need for compulsive assessment and therapy. Emerg Med Rep 5:45–52

Insall JN (1984) Meniscectomy. In: Insall JN (ed) Surgery of the knee. Churchill Livingstone, New York, p 828

Ishikawa H, Weba Y, Yonezawa T (1989) Osteochondritis dissecans of the shoulder in a tennis player. Am J Sports Med 16(5):547–550

Jackson DW, Jennings LD, Maywood RM, Berger PE (1988) MRI of the Knee. Am J Sports Med 16(1):29–38

Kaye J, Bohne W (1977) A radiographic study of the ligamentous anatomy of the ankle. Radiology 125:659–667

Kaye JJ, Nance EP (1987) Pain in the Athlete's Knee. Clin Sports Med 6(4):873–883

Kean DM, Worthington BS, Preston BJ, Roebuck EJ, McKim-Thomas H, Hawkes RC, Holland GN, Moore WS (1983) Nuclear MR imaging of the Knee: Examples of Normal Anatomy and Pathology. Br J Radiol 56:355–364

Keene J, Lash E, Fisher D, De Smet A (1989) MR imaging of Achilles tendon ruptures. Am J Sports Med 17(3):333–337

Kelikian H, Kelikian A (1985) Disorders of the ankle. WB Saunders, Philadelphia

Kennedy JC, Hawkins R, Krissoff WB (1978) Orthopedic manifestations of swimming. Am J Sports Med 6:309

Kieft GJ, Bloem JL, Oberman WR, Verbout AJ, Rozing P, Doornbos J (1986) MR imaging of the normal shoulder. Radiology 159:741–745

Kieft GJ, Sartoris DJ, Bloem JL, Hajek PC, Baker LL, Resnick D, Obermann WR, Rozing P, Doornbos J (1987) MR imaging of glenohumeral joint diseases. Skeletal Radiol 16:285–290

Kieft GJ, Bloem JL, Rozing PM (1988a) Rotator cuff impingement syndrome: MR imaging. Radiology 166:211–214

Kieft GJ, Bloem JL, Rozing PM (1988b) MR Imaging of recurrent anterior dislocation of the shoulder: comparison with CT, Arthrography. AJR 150:1083–1087

Kneeland JB, Carrera GF, Middleton WD, Campagna NF, Ryan LM, Jesmanowicz A, Froncicz W, Hyde JS (1986) Rotator cuff tears: preliminary application of high resolution MR imaging with counter rotating current loop-gap resonators. Radiology 160:695–699

Kneeland JB, Middleton WD, Carrera GF, Zeuge RC, Jesmanowicz A, Froncisz W, Hyde JS (1987) MR imaging of the shoulder: diagnosis of rotator cuff tears. AJR 149:333–347

Kneeland JB, Macrandar S, Middleton WD (1988) Pictorial essay MR imaging of the normal ankle: correlation with anatomic sections. AJR 151:117–126

Koenig H, Lucas D, Meissner R (1986) The wrist: a preliminary report on high resolution MR imaging. Radiology 160:463–467

Koenig H, Sauter R, Deimling M, Vogt M (1987) Cartilage Disorders: Comparison of Spin-Echo, CHESS and FLASH Sequence MR Images. Radiology 164:753–758

Kottal RA, Vogler II JB, Matamoros A, Alexander AH, Cookson JL (1987) Pigmented Villonodular Synovitis: A Report of MR Imaging of Two Cases. Radiology 163:551–553

Kujala VM, Kvist M, Osterman K (1986) Knee Injuries in athletes. Sports Med 3:447–460

Kulkarni MV, Patton JA, Price RR (1986) Technical Considerations for the use of Surface Coils in MRI. AJR 147:373–378

Kulund DN (1988) The knee. In: Kulund JN (ed) The injured athlete. Lippincott, Philadelphia, p 603

Lang P, Jergesen HE, Mosely ME (1988) Avascular Necrosis of the Femoral Head: High Field Strength MR Imaging with Histologic Correlation. Radiology 169:517–524

Larson RL, Singer KM (1985) The Knee, Clinics in Sports Medicine. WB Saunders, Philadelphia, p 401

Lee KR, Glendon GC, Neff JR, Arnett GR, Murphey MD (1987) Cystic Masses of the Knee: Arthrographic and CT Evaluation. AJR 148:329–334

Lee JK, Yao L (1988) Stress fractures: MR imaging. Radiology 169:217–220

Lee JK, Yao L, Phelps CT, Wirth CR, Czajka J, Lozman J (1988) Anterior Cruciate Ligament Tears: MRI Compared with Arthroscopy and Clinical Tests. Radiology 166:861–864

Lehner KB, Rechl HP, Omeinwieser JK, Heuck AF, Lukas HP, Kohl HP (1989) Structure, Function and Degeneration with MR imaging in vitro. Radiology 170:495–599

Levinsohn EM, Baker BE (1980) Prearthrotomy Diagnostic evaluation of the Knee: Review of 100 cases diagnosed by Arthrography and Arthroscopy. AJR 134:107–111

Li DKB, Adams ME, McConkey JP (1986) MR Imaging of the Ligaments and Menisci of the Knee. Radiol Clin North Am 24:209–227

Li KC, Henkelman RM, Poon PY, Rubenstein J (1984) MR Imaging of the Normal Knee. J Comput Assist Tomogr 8:1147–1154

Lodwick GS, Rosenthal DI, Kattapuram SV, Hudson TM (1987) Fatigue and Insufficiency fractures. J Med Imaging 1–9

Lotysch M (1986) Scrutinizing Knee Joints: MRI Offers New Insight. Diagn Imaging 8:90–94

Lotysch M, Mink J, Crues JY, Schwartz SA (1986) MR imaging in the detection of meniscal injuries. Magn Reson Imaging 4:94

Lynch TCP, Crues III JV, Morgan FW, Sheeman WE, Harter LP, Ryu R (1989) Bone abnormalities of the knee: prevalence and significance at MR imaging. Radiology 171:761–766

Mack LA, Matsen FA (1985) US evaluation of the rotator cuff. Radiology 157:205–209

Maldague B, Malghem J (1978) Chondromalacie de la rotule, apport de la radiologie. Acta Orthop Belg 44:21–40

Malghem J, Maldague B (1989) Depth insufficiency of the proximal trochlear groove on lateral radiographs of the knee: relation to patellar dislocation. Radiology 170:507–510

Manco LG, Lozman J, Coleman ND, Kavanaugh JH, Bilfield B, Dougherty J (1987) Noninvasive Evaluation of Knee Meniscal Tears: Preliminary Comparison of MR Imaging and CT. Radiology 163:727–730

Mandelbaum BR, Finerman GAM, Reicher MA, Hartzman S, Bassett LW, Gold RH, Rauschning W, Dorey F (1986) MR Imaging as a Tool for Evaluation of Traumatic Knee Injuries. Am J Sports Med 14:361–370

Mayfield GW (1977) Popliteus tendon tenosynovitis. Am J Sports Med 5:31

McCue FC, Baugher WH, Bourland WL, Brunet ME (1979a) Hand injuries in athletes. Surg Rounds 1:8

McCue FC, Baugher WH, Kulund DN, Gieck JH (1979b) Hand and wrist injuries in the athlete. Am J Sports Med 7:275

McGlade CT, Bassett LW (1989) The hip. In: Bassett LW et al. (eds) Atlas of the musculoskeletal system. Dunitz, London, p 415

McGuinty JB (1984) Arthroscopy of the knee. In: Insall JN (ed) Surgery of the Knee. Churchill Livingstone, New York, p 828

Mesgarzadeh M, Revesz G, Bonakdarpour A (1985) The effects on the Medical Metal Implants by Magnetic Fields of MR Imaging. Skeletal Radiol 14:205–206

Mesgarzadeh M, Sapega AA, Bonakdarpour A, Revesz G, Moyer RA, Maurer AH, Alburger PD (1987) Osteochondritis Dissecans: Analysis of Mechanical Stability with Radiography, Scintigraphy and MRI. Radiology 165:775–780

Meurmann K (1981) Less common stress fractures of the foot. Br J Radiol 54:1–7

Middleton WD, Kneeland JB, Carrera GF, Cates JD, Kellman GM, Campagna NG, Jesmanowicz A, Froncisz W, Hyde JS (1987a) High-resolution MR imaging of the normal rotator cuff. AJR 148:559–564

Middleton WD, Kneeland JB, Kellman GM, Cates JD, Sanger JR, Jesmanowicz A, Froncisz W, Hyde JS (1987b) MR Imaging of the carpal tunnel: Normal anatomy and preliminary findings in the carpal tunnel syndrome. AJR 148:307–316

Middleton WD, Macrander S, Kneeland JB, Froncisz W, Hyde JS, Jesmanowicz A (1987c) MR imaging of the normal elbow: anatomic correlation. AJR 149:543–547

Mink JH, Deutsch AL (1989) Occult Cartilage and Bone Injuries of the Knee: Detection, Classification and Assessment with MR Imaging. Radiology 170:823–829

Mink JH, Reicher MA, Crues III JV (1987) MRI of the knee. Raven Press, New York, p 178

Mink JH, Levy T, Crues III JV (1988) Tears of the anterior cruciate ligament and menisci of the knee: MR evaluation. Radiology 167:769–774

Mitchell DG (1989) Using MR Imaging to Probe the Pathophysiology of Osteonecrosis. Radiology 171:25–26

Mitchell DG, Rao VM, Dalinka M (1986) Hematopoietic and Fatty Bone Marrow Distribution in the Normal and Ischemic hip: New Observations with 1,5 T MR Imaging. Radiology 161:199–202

Mitchell DG, Rao V, Dalinka M, Spritzer CE, Alavi A, Steinberg ME, Fallon M, Kressel HY (1987a) Femoral Head Avascular Necrosis: Correlation of MRI, Radiographic Staging, Radionuclide Imaging and Clinical Findings. Radiology 162:709–715

Mitchell DG, Burk DL, Vinitski S, Rifkin MD (1978b) The Biophysical Basis of Tissue Contrast in Extracranial MRI. AJR 149:831–837

Mitchell MJ, Causey G, Berthoty DP (1988) Peribursal fat plane of the shoulder: anatomic study and clinical experiences. Radiology 168:699–704

Moon KL, Genant HK, Helms CA, Chafetz N, Crooks LE, Kaufman L (1983) Musculoskeletal Applications of NMR. Radiology 147:161–171

Morti R (1977) Dislocation of peroneal tendon. Am J Sports Med 5(1):19–22

Müller W (1982) Das Knie. Springer-Verlag, Berlin Heidelberg New York, p 122

Murphy WA, Siegel MJ (1977) Elbow Fat Pads with with new signs and extended differential diagnosis. Radiology 124:659–665

Muwanga C, Heller M, Quinton D (1986) Grade III injuries of the lateral ligaments of the ankle: the incidence and a simple stress test. Arch Emerg Med 3:247–251

Negendank WG, Fernandez-Madrid FR, Heilbrun LK (1989) MRI of meniscal degeneration in asymptomatic knees: evidence of a relation to clinical tears. Book of abstracts, 8th Annual Meeting SMRM, 1:279

Nicholas JA, Hershman EB (1986) The lower extremity and spine in Sports Medicine. Mosby, St Louis, p 1085

Niedermann B (1981) Rupture of the lateral ligaments of the ankle: Operation or plaster cast? Acta Orthop Scand 52:579–587

Noto AM, Cheung Y, Rosenberg ZS, Norman A, Leeds NE (1989) MR Imaging of the ankle: normal variants. Radiology 170:121–124

Odensten M, Hamberg P, Nordin M, Lysholm J, Gillquist J (1985) Surgical or Conservative treatment of the acutely torn anterior cruciate ligament: a randomized study with short-term follow-up observations. Clin Orthop 198:87–93

O'Neill D, Micheli L (1989) Recognizing and preventing overuse injuries in young athletes. J Musculoskel Med 6(2):106–126

Oppelt A, Graumann R, Barfuss H, Fischer H, Hartl W, Schajor W (1986) FISP, A New Fast MRI Sequence. Electromedica 54:15–18

Orava S (1978) Iliotibial tract friction syndrome in athletes. Br J Sports Med 12:69

Pappas AM (1986) Injuries of the Upper Extremities. In: Vinger PF, Hoerner EF (eds) Sports Injuries. PSG Littleton, Massachusetts, p 433

Perlmann M, Leveille D, Deleonibus J (1987) Inversion lateral ankle trauma: differential diagnosis, review of the literature and prospective study. J Foot Surg 26:95–121

Pierre-Jerome C, Shahabpour M, Achten E, Luypaert R, Osteaux M (1988) Comparative wrist MRI, its role in the differential diagnosis of the painful wrist. 7th Annu Meeting SMRM. Book of works in progress, 26

Pierre-Jerome C, Shahabpour M, Luypaert R, Osteaux M (1989) Bilateral MR imaging of the wrist: evaluation of pre- and postoperative carpal tunnel syndrome. Radiology 173(P):504

Pollack MS, Dalinka MK, Kressel HY, Lotke PA, Spritzer CE (1987) MR Imaging in the Evaluation of Suspected Osteonecrosis of the Knee. Skeletal Radiol 16:121–127

Quinn SF, Murray WT, Clark RA, Cochran CF (1987) Achilles tendon: MR imaging at 1.5 T. Radiology 164:767–770

Quinn SF, Muus C, Sara A (1988) Meniscal tears: pathologic correlation with MR imaging. Radiology 166:580

Reicher MA, Bassett LW, Gold RH (1985a) High-Resolution MR Imaging of the Knee Joint: Pathologic Correlations. AJR 145:903–909

Reicher MA, Rauschning W, Gold RH, Bassett LW, Lufkin RB, Glen W (1985b) High-Resolution MR Imaging of the Knee Joint: Normal Anatomy. AJR 145:895–902

Reicher MA, Hartzman S, Duckwiler GR, Bassett LW, Anderson LJ, Gold RH (1986) Meniscal Injuries: Detection Using MR Imaging. Radiology 159:753–757

Reicher MA, Hartzman S, Bassett LW, Mandelbaum B, Duckwiler G, Gold RH (1987) MR Imaging of the Knee, Part I, Traumatic Disorders. Radiology 162:547–551

Reinus WR, Conway WF, Totty WG, Gilula LR, Murphy WA, Siegel BA, Weeks PM, Young VL, Manske PR (1986) Carpal avascular necrosis: MR Imaging. Radiology 160:689–693

Reis ND, Lanir A, Benmair J, Hadar H (1985) MRI in orthopedic surgery – a glimpse into the future. J Bone Joint Surg 67:659–664

Reiser MF, Bongartz G, Erlemann R, Strobel M, Pauly T, Gaebert K, Stoeber U, Peters P (1988) MR in cartilaginous lesions of the knee joint with three-dimensional gradient-echo imaging. Skeletal Radiol 17:465–471

Richardson ML (1986) MRI of the Musculoskeletal System. Radiol Clinics of North America, 24, 2, WB Saunders, Philadelphia, p 335

Resnick D (1989) Bone and Joint Imaging. WB Saunders, Philadelphia, p 1330

Retting A, Shelbourne K, Beltz H, Arfken P (1987) Radiographic evaluation of foot and ankle injuries in the athlete. Clin Sports Med 6(4):905–919

Roels J, Martens M, Mulier JC, Burssens A (1978) Patellar Tendinitis. Am J Sports Med 6:362–368

Roger B, Christel P, Paix D, Saillant G, Cabanis E (1987) Imagerie par résonance magnétique des lésions de l'aponécrose plantaire. J Radiol 68(12):749–753

Rosenberg Z, Feldmann F, Swingson R, Price G (1987) Peroneal tendon injury associated with calcaneal fractures, CT findings. AJR 149:125–129

Sarrat P, Acquaviva PC, Lafforgue P, Zakarian H, Lopez M, Bouscarle B (1988) L'algodystrophie de la tête fémorale. J Radiol 69(8–9):495–500

Sartoris DJ, Resnick D (1988) Imaging the painful shoulder: what studies to order. J Musculoskel Med 5:21–35

Schneider RC, Kennedy JC, Plant ML (1985) Sports Injuries: Mechanisms, Prevention and Treatment. Williams & Wilkins, Baltimore, p 877

Seeger LL (1989) MRI of the shoulder. Clin Orthop 244:48–59

Seeger LL, Ruszkowski JT, Bassett LW, Kay SP, Kahmann RD, Ellman H (1987) MR imaging of the normal shoulder: anatomic correlation. AJR 148:83–91

Seeger LL, Gold RH, Bassett LW (1988a) Shoulder impingement syndrome: MR findings in 53 shoulders. AJR 150:343–347

Seeger LL, Gold RH, Bassett LW (1988b) Shoulder instability: evaluation with MR imaging. Radiology 168:695–697

Seiler JG, Christie MJ, Homra L (1989) Correlation of the findings of MRI with those of bone biopsy in patients who have stage I or II ischemic necrosis of the femoral head. J Bone Joint Surg 71A:28–32

Shahabpour M, Osteaux M (1989) Imagerie par Résonance Magnétique. In: Sintzoff S (ed) Imagerie du genou. Masson, Paris, p 133

Shahabpour M, Achten E, Vaes P, Mattheus R, Luypaert R, Osteaux M (1988a) The anterior talo-fibular (ATF) ligament of the ankle: comparison of classical spin echo imaging and fast 3D MRI. Br J Radiol 61:728

Shahabpour M, Osteaux M, Achten E, Mattheus R, Luypaert R (1988b) 3D-analysis of fast high-field MRI: clinical application to the knee joint. Br J Radiol 61:728

Shahabpour M, Van Roy P, Osteaux M, Luypaert R (1988c) MRI of the Knee. J Belge Radiol 71:241–253

Shahabpour M, Van Roy P, Van Cauteren M, Osteaux M, Casteleyn PP (1989a) Three-dimensional analysis of the knee joint with fast MRI: correlation with arthroscopic findings. AM J Sports Med 17:5

Shahabpour M, Van Roy P, Van Cauteren M, Osteaux M (1989b) Patello- and tibiofemoral cartilaginous lesions: multiplanar analysis with 3D MR imaging. Radiology 173(P):171

Shahabpour M, Van Cauteren M, Handelberg F (1989c) MRI of chondral lesions of the knee: a comparative study of 2D spin echo and 3D gradient echo imaging. Book of Abstracts, 8th Annual Meeting SMRM 1:277

Shahriaree H (1985) Chondromalacia. Contemp Orthop 11:27–39

Sherman OH, Fox JM, Snyder SJ, Del Pizzo W, Friedman MJ, Ferkel RD, Lawley MJ (1986) Arthroscopy, no problem surgery: an analysis of complications in 2640 cases. J Bone Joint Surg 68:256–265

Sloth C, Lundgrenjust S (1989) The apical oblique radiograph in examination of acute shoulder trauma. Eur J Radiol 9:147–151

Soudry M, Lanir A, Angel D, Roffman M, Kaplan N, Mendes DG (1986) Anatomy of the Normal Knee as seen by MR Imaging. J Bone Joint Surg 68-B:117–120

Spritzer CE, Vogler JB, Martinez S, Garrett WE, Johnson GA, McNamara MJ, Lohne SJ, Herfkens RJ (1988) MR imaging of the knee: preliminary results with a 3DFT GRASS pulse sequence. AJR 150:597–604

Stafford SA, Rosenthal DI, Gebhardt MC, Brady TJ, Scott JA (1986) MRI in Stress Fracture. AJR 147:553–556

Stark HH, Jobe FW, Boyes JH, Ashworth CR (1977) Fracture of the hook of the hamate in athletes. J Bone Joint Surg 59A:575

Stoller DW (1988) Fast MR improves imaging of musculoskeletal system. Diagn Imaging 2:98–197

Stoller DW, Martin C, Crues III JV, Kaplan L, Mink JH (1987) Meniscal Tears: Pathologic Correlation with MR Imaging. Radiology 163:731–735

Sundaram M, McGuire MH, Fletcher J, Wolverson MK, Heiberg E, Shields JB (1986) MR Imaging of Lesions of Synovial Origin. Skeletal Radiol 15:110–116

Talersnik J, Kelly PJ (1966) The extraosseous and intraosseous blood supply of the scaphoid bone. J Bone Joint Surg 48A:1125

Teng MM, Murphy WA, Gilula LA (1984) Elbow Arthrography: a reassessment of the technique. Radiology 153:611–613

Thijn CJP (1982) Accuracy of Double-Contrast Arthrography and Arthroscopy of the Knee Joint. Skeletal Radiol 8:187–192

Turner DA, Prodromos CC (1986) MR imaging of knee injuries. Semin Ultrasound 7:339–350

Turner DA, Prodromos CC, Petasnik JP, Clark JW (1985) Acute Injury of the Ligaments of the Knee: MR evaluation. Radiology 154:717–722

Turner DA, Templeton AC, Selzer PM (1989) Femoral capital osteonecrosis, MR finding of diffuse marrow abnormalities wihout focal lesions. Radiology 171:135–140

Tyrrell RL, Gluckert K, Pathria M, Modic HT (1988) Fast three-dimensional MRI of the knee: comparison with arthroscopy. Radiology 166:865–872

Vaes P, Shahabpour M, Van Cauteren M, Osteaux M (1989) 3D analysis of injured lateral ligaments of the ankle with FISP MR imaging. Radiology 173(P):272

Van Cauteren M, Shahabpour M, Vaes P, Osteaux M (1989) Correct spatial angle determination using MRI. Book of Abstract, 8th Annual Meeting SMRM. 2:931

Vinger PF, Hoerner EF (1986) Sports injuries. The unexpected epidemic. PSG Littleton, Massachusetts, p433

Weiss KL, Beltran J, Shamam OM, Stilla RF, Levey ML (1986a) High field MR surface coil imaging of the hand and wrist, part I, Normal anatomy. Radiology 160:143–146

Weiss KL, Beltran J, Lubbers LM (1986b) High field MR surface coil imaging of the hand and wrist, part II, Pathologic correlations and clinical relevance. Radiology 160:147–152

Williams A, Evans R, Shirley PD (1989) Imaging of sports injuries. Baillère Tindall. WB Saunders, London, p 182

Wilson E, Katz F (1969) Stress fractures, analysis of 250 consecutive cases. Radiology 92:481–486

Wojtys E, Wilson M, Buckwalter K, Braunstein E, Martel W (1987) MR Imaging of Knee Hyaline Cartilage and Intraarticular Pathology. Am J Sports Med 15:455–463

Yao L, Lee JK (1988) Occult intraosseous fracture: detection with MR Imaging. Radiology 167:749–751

Yulish BS, Lieberman JM, Newman AJ, Bryan PJ, Mulopulos GP, Modic MT (1987a) Juvenile Rheumatoid Arthritis: Assessment with MR Imaging. Radiology 165:149–152

Yulish BS, Montanez J, Goodfellow DB, Bryan PJ, Mulopulos GP, Modic MT (1987b) Chondromalacia Patellae: assessment with MRI. Radiology 164:763–766

Yulish BS, Mulopulos GP, Goodfellow DB, Bryan PJ, Modic TM, Dollinger BM (1987c) MR Imaging of Osteochrondral Lesions of Talus. J Comput Assist Tomogr 11 (2):296–301

Zarins B, Boyle J (1986) Knee ligament injuries. In: Nicholas JA, Hershman EB (1986) The lower extremity and spine in sports medicine. Mosby, St Louis, p 1085

8 The Spine

T. Buisseret, T. Stadnik, and E. Verhaeve

8.1 Introduction

Spine injuries account for only 5%–15% of all sport-related traumata, but they represent a high proportion of the most devastating injuries in sports medicine. An increased number of severe spine injuries with permanent neurological damage have been reported in a wide variety of sports activities including automobile and motorcycle racing, diving, parachuting, trampolining, gymnastics, and horse riding.

With a superior tissue differentiation and the major advantage of multiplanar imaging, MRI is ideally suited for these heavily traumatized patients, for whom an early and accurate diagnosis is mandatory. MRI is the only imaging modality that allows simultaneous visualization of the bones, ligaments, intervertebral disks, and spinal cord without contrast injection, myelography, or hazardous repositionning of the patient. It should be stressed that MRI is totally noninvasive and is performed on an immobilized patient lying supine in the machine. Sagittal imaging allows a clear anatomic overview of the whole spine, including the spinal canal and the spinal cord, and permits a distinction between stable and unstable fractures (Athanosoulis and Lee 1987; Chakeres et al. 1987).

MRI has also proven extremely valuable in the assessment of minor injuries often involving the intervertebral disks. Gymnasts particularly are prone to intervertebral disk disease due to constant pressure and traction forces on the various spine segments; tennis and volleyball are other high-risk activities with sudden bending and stretching of spine segments. Swinging movements with rackets or golf clubs may also lead to disk disease and radicular compression. Disk lesions are not always conspicuous at once; the cumulative effect of repeated, minor injuries without any visible changes on conventional radiographs can lead to significant symptoms developing late in the athlete's carreer or even after retirement. By its greater tissue sensitivity, MRI may very well detect early changes in the substance of the disk itself (Grenier et al. 1987; Lipson and Muir 1981; Pech and Haughton 1985).

8.2 Technical Factors

Unlike CT and conventional roentgenograms, which are dependent on information related to electron density, proton MR signals depend on the

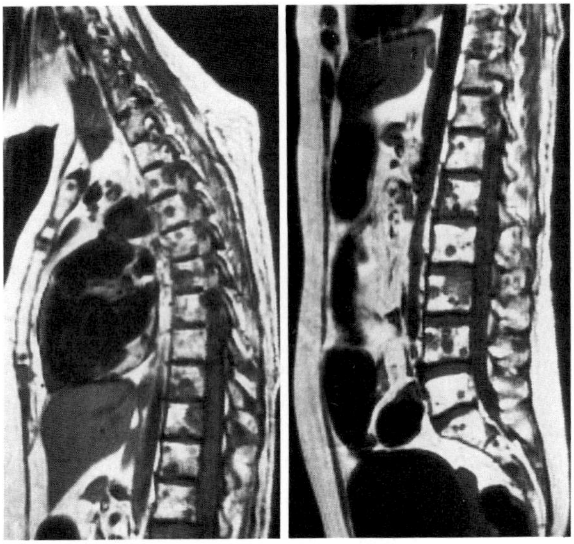

Fig. 8.1. Whole spine, body coil (SE 600/15, 4 mm). Midsagittal T_1-weighted images providing an overview of the whole spine. Metastatic replacement of bone marrow, involving several vertebral bodies and the sternum

hydrogen content of the examined tissues and on their relative T_1 and T_2 relaxation characteristics. The much greater tissue sensitivity and contrast resolution thus allows simultaneous visualization of osseous and nonosseous lesions (Stark and Bradley 1988).

The patient is laid supine in the gantry or magnet; whenever possible a surface coil is placed under the involved spine segment. This is a small receiver antenna providing a much higher signal-to-noise ratio and thus a higher resolution and improved diagnostic accuracy. Surface coils, however, have a field of view of only 20–25 cm and thus examine only a part of the spine rather than its totality. Obviously these coils cannot be used when the specific level cannot be clinically determined, or when a diffuse involvement of the spine is suspected, as in metastatic disease (Fig. 8.1) (McArdle et al. 1986).

8.3 MR Anatomy

A T_1-weighted spin-echo sequence provides an excellent anatomic approach of the spine; here both repetition time (TR 400–600 ms) and echo time (TE 15–40 ms) are short. As the imaging time of a sequence is determined largely by the repetition time, this sequence does not last longer than 5–12 min, depending on the number of acquisitions required.

The vertebral bone marrow on a T_1-weighted sequence shows intermediate signal intensity. The vertebral bodies are bordered by a rim of low intensity

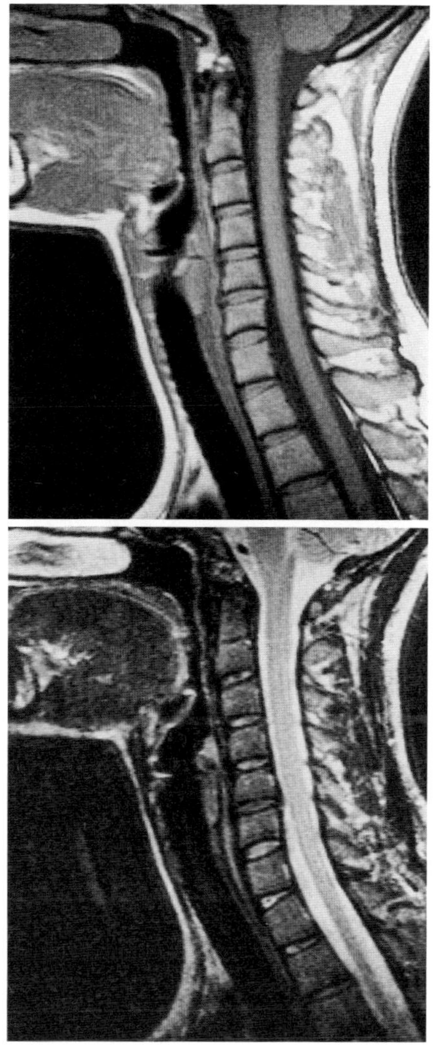
a

b

Fig. 8.2. a, b. Cervical spine, surface coil. Normal anatomy, spin-echo sequences. **a** Midsagittal T_1-weighted image (SE 600/15, 3 mm). **b** Midsagittal T_2-weighted image (SE 2000/90, 3 mm). Note the myelographic effect with the bright signal of the cerebrospinal fluid surrounding the darker spinal cord

corresponding to cortical bone, which produces little or no signal due to low hydrogen content. The intervertebral disks are biconvex and produce a homogeneous signal of intermediate intensity, slightly lower than that of bone marrow and similar to the signal of muscle tissue. The disks should be clearly outlined and should not protrude behind the posterior border of the adjacent vertebral bodies (Fig. 8.2a). The spinal cord, appearing white, is surrounded by cerebrospinal fluid with low signal intensity and thus dark on this short TR, short TE spin-echo sequence. Epidural fat, on the other hand, has a very high and bright signal on both T_1- and T_2-weighted spin-echo sequences, providing a very useful contrast to the exiting nerve roots. With a slice thickness of 3–

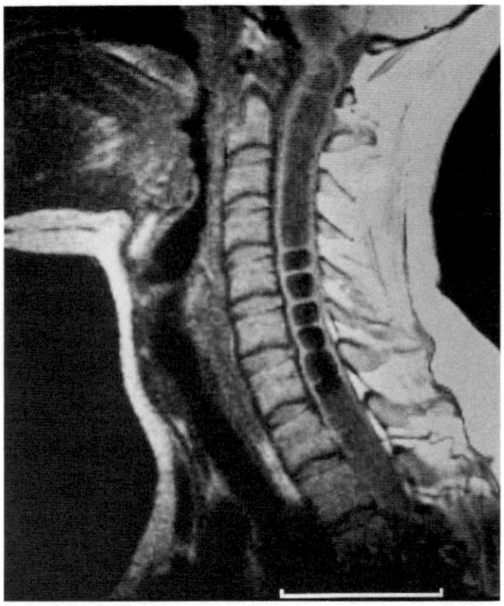

Fig. 8.3. Cervical spine, surface coil (SE 600/15, 3 mm). Posttraumatic syrinx. The centromedullar cavity is readily outlined on a T_1-weighted spin-echo sequence

4 mm the exiting nerve roots are indeed well visualized at the level of the radicular foramina on both axial and sagittal images.

In most traumatic instances a T_1-weighted spin-echo sequence is sufficient to evaluate bone and soft tissue injuries. The images are generally obtained in a sagittal plane; axial images may be necessary to evaluate disk herniations and radicular lesions. A traumatic syrinx is easily detected as a centromedullar linear lucency (Fig. 8.3). A T_2-weighted spin-echo sequence with a long repetition time (TR 1500–2300 ms) and long echo times (TE 70–100 ms) may be necessary to evaluate more precisely medullar lesions resulting from contusion or ischemia. On these T_2-weighted images the vertebral bodies still have a low to intermediate signal intensity and a rim of dark cortical bone; the spinal cord, on the other hand, appears black and is surrounded by cerebrospinal fluid with a very bright signal due to a long T_2 relaxation time. The images are quite similar to myelography, where the dark spinal cord is outlined by the intrathecal contrast medium (Fig. 8.2b) (Modic et al. 1986a, b; Tarr et al. 1987).

8.4 Injuries to the Spine

8.4.1 Cervical Spine

Bony injuries to the spine can affect all cervical vertebrae; lesions of C1 (Jefferson's fractures) and of C2 usually result from direct compression forces and are observed in diving accidents and high-speed traumata. They usually cause severe neck pain but are seldom associated with neurological lesions. These fractures are usually well documented by axial CT scanning, clearly outlining the disruption of the bony ring of the atlas. Magnetic resonance can disclose associated lesions to the transverse ligament.

The so-called hangman's fracture involves the pedicles of C2 and is common in high-velocity injuries, but it also occurs with increased frequency in football, ice hockey, gymnastics, and aquatic and equestrian sports. The intervertebral disk at the C2–C3 level may be damaged and produce neurological injury. While these lesions are usually well documented by anteroposterior and lateral roentgenograms taken in the emergency department, CT or MRI are important diagnostic tools in evaluating the spinal cord and in establishing the patency of the spinal canal (Quencer et al. 1986).

Fractures of the lower cervical bodies usually result from a combination of compression and flexion forces. According to Cloward's classification, these may be divided into four types. Type 1 injuries are common and show a comminution of the anterosuperior or -inferior margin of the cervical body, often producing kind of a teardrop fracture. Type 2 injuries produce a more extensive comminution of the cervical body but still without comminution of the posterior wall. Type 3 fractures include extensive comminution of the anterior cortex but leave the posterior cortex intact, although it may be displaced posteriorly. Major comminution of the vertebral body with disruption of the posterior wall characterizes type 4 lesions; in this type of fractures a retropulsion of bone fragments into the spinal canal may be observed. The neural function may be compromised in type 3 and 4 fractures by narrowing of the spinal canal and compression of the medullar cord. Again, sagittal MRI allows an easy and rapid evaluation of possible medullar compression. However, small osseous fragments, usually composed of cortical bone and a low or absent signal on MRI examination. In these instances axial CT slices with a slice thickness of 1–2 mm should be advocated to localize these bony chips (Modic et al. 1986a; Tarr et al. 1987).

So-called clay shoveler's fractures follow severe flexion movements with fractures of the spinous processes. When affecting the low cervical vertebrae, especially C6 and C7, these fractures are frequently misdiagnosed on the initial roentgenograms because visualization of the area is poor on plain films due to the superposition of the shoulder densities, which is often a problem with large-shouldered athletes. MRI offers a useful alternative to CT in these cases, with a better approach of the spinous processes in a sagittal plane.

Facet joint injuries include subluxation or dislocation of the facet joint, which may be unilateral or bilateral. Especially the bilateral dislocations are accompanied by significant disruption of the adjacent disk and supporting ligamentous structures, with a significant displacement of one cervical vertebra on the other, possibly leading to medullar compression by severe angulation of the spinal cord. Sagittal imaging provides the clue to diagnosis, evaluating the alignment of the vertebral bodies, the degree of disk herniation, and the extent of angulation and compression of the spinal cord (Fig. 8.4) (Quencer et al. 1986).

Ligamentous lesions may occur in combination with cervical fractures or alone; acute flexion injuries account for posterior ligamentous disruption.

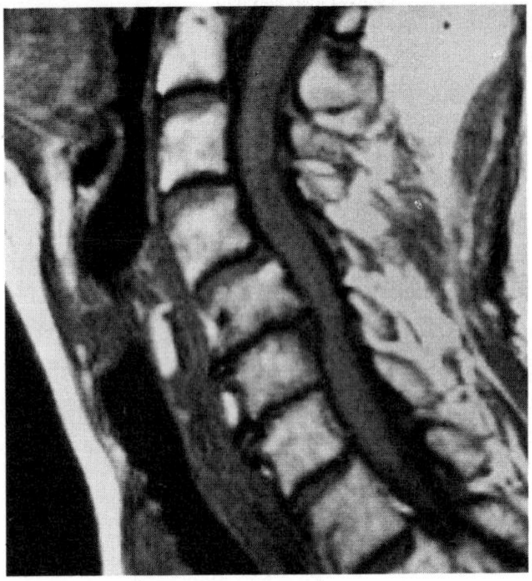

Fig. 8.4. Cervical spine, surface coil (SE 600/15, 3 mm). Traumatic subluxation; post-traumatic degenerative changes involving C5 and C6 with a narrowing of the spinal canal and a local medullar compression

Usually the interspinous ligaments are involved first, with an obvious widening of the interspinous gap on lateral plain films. Flexion and extension views should be taken to appreciate the extent of the lesion. In more severe flexion traumata, as observed in diving accidents, disruption of the ligamenta flava and posterior longitudinal ligament may occur simultaneously. In these cases the disk itself may be injured, with a subsequent degeneration of the disk tissue and a narrowing of the disk space. Associated bone injury along the anterior border of the vertebral body may be present, but the anterior longitudinal ligament is seldom involved (Fig. 8.5). Hyperextension movements may injure the anterior longitudinal ligament, the cervical disk, and, less commonly, the posterior longitudinal ligament.

A combination of acute flexion and extension occurs in the so-called whiplash injuries so typical of automobile crashes, but they also observed in various sports, including gymnastics, rugby, and diving. Whiplash injuries may involve all neck tissues, including trachea, esophagus, ligaments, cervical disks, and muscles. Hematomas can be discerned among the torn muscle fibers on MR images. The longitudinal anterior and posterior ligaments can also be identified, but ligamentous tears are more difficult to objectivate because up to now MRI is essentially static, and dynamic flexion and extension movements cannot yet be registered.

Disk lesions are very common and have been observed in all kinds of sport activities; most are unilateral, involve one segment, and result in radicular symptoms due to compression of the nerve root. These posterolateral disk herniations are diagnosed either by axial CT scanning or MRI. With MRI parasagittal and axial images are mandatory; oblique images perpendicular to the long axis of the neural foramina can be obtained for optimal visualization of the nerve roots. In middle-aged and elderly persons the nerve root compression is accentuated in most instances by preexisting uncarthrosis and a supplementary narrowing of the neural foramina. Central disk

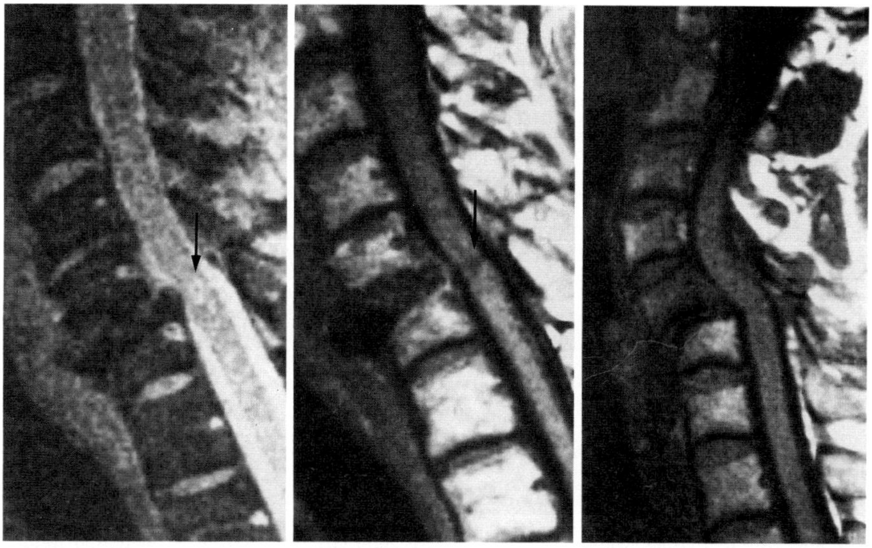

a b c

Fig. 8.5 a–c. Cervical spine, surface coil. **a** Midsagittal T_1-weighted spin-echo sequence (SE 600/15, 3 mm). Traumatic luxation with major disk herniation C 5–C 6 and an angulation of the spinal cord. **b, c** Same patient; midsagittal T_1-weighted (**b**) and T_2-weighted (**c**) spin-echo sequences. Surgical reposition with an anterior bone graft. Note the pathological signal in the spinal cord (*arrows*) representing ischemic tissue due to compression of the anterior spinal artery

herniation is rare but may cause a contusion to the spinal cord; all these lesions can be aggravated by a preexisting, congenital or arthrotic narrowness of the spinal canal, which is clearly evaluated by a sagittal overview of the cervical spine.

Spinal cord lesions are encountered in contact sports such as rugby, football, boxing, karate and in high-speed traumata. When they are not associated with significant osseous lesions, they seem to be due to contusion or possibly ischemia resulting from impairment of the anterior spinal artery. T_2-weighted images may show a focal area of high signal intensity at the site of medullar contusion (Fig. 8.6).

It is still difficult to objectivate isolated nerve root injuries. If the neural foramen is already narrowed by cervical spondylosis, a contusion of the exiting nerve may result from even minor traumata. If the narrowing of the neural foramina is obvious on parasagittal or oblique T_1-weighted images, the signal and aspect of the nerve are unchanged. Avulsion injuries of the cervical nerve roots, sustained during high-velocity accidents, cannot unequivocally be demonstrated by MRI. When this is suspected clinically, diagnostic is only a CT scan taken after a cervical myelogram and showing a characteristic pooling of contrast dye around the sleeve of the avulsed nerve root.

8.6 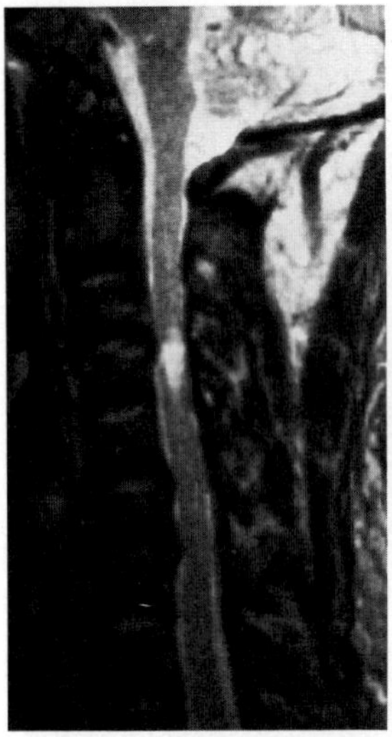

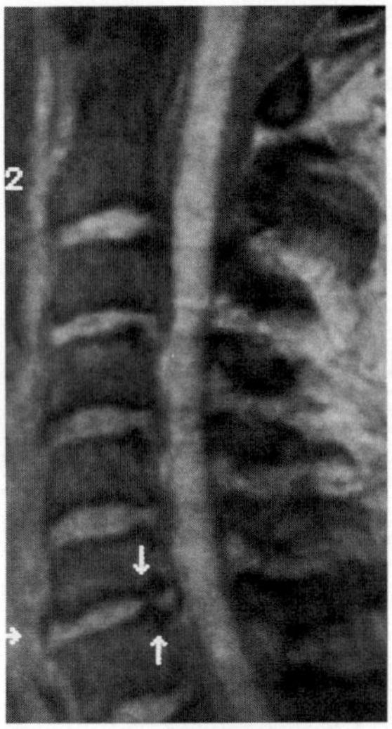

8.7

Fig. 8.6. Cervical spine, spin-echo sequence (SE 2000/90). Midsagittal T_2-weighted image. Stenosis of the spinal canal resulting from degenerative, arthrotic ventral and dorsal ridging. Obvious obliteration of the anterior epidural space and compression of the spinal cord with a focal intramedullar lesion with high signal intensity

Fig. 8.7. Cervical spine, gradient-echo sequence (70° flip angle, 300/12). Same patient as in Fig. 8.6. The faster gradient-echo sequence provides a good definition of osseous changes: note the ostophytic appositions (*arrows*). The spinal cord, however, is difficult to evaluate on this sequence; the intramedullar lesion clearly visualized on Fig. 8.6 cannot be detected

8.4.2 Thoracic Spine

Most injuries to the thoracic spine concern wedge-compression fractures resulting from high-velocity sports such as automobile racing or downhill skiing. Plain X-ray documents frequently reveal a comminution or wedging deformity of the involved thoracic vertebral body, however these give only a partial picture of the the extent of trauma due to a difficult visualization of the posterior wall and the posterior facet joints. Lateral tomograms and CT scans may be necessary for better evaluation of the lesions and the spinal canal. Sagittal MRI imaging can replace both techniques advantageously by depicting the osseous lesions as well as their possible impingement on the spinal cord.

Thoracic disk lesions are rare, but they can result from a disruption of the superior endplate of a compressed vertebral body. The objectivation of thoracic disk herniations has always been difficult. Most patients present with vague symptoms of myelopathy and with few clinical signs and no absolute neurological or radiographic evidence to indicate which segment of the thoracic spin is involved. Axial CT sections with a slice thickness of 4–5 mm cannot possibly investigate the whole thoracic spine. Myelography is more efficient, allowing a sagittal overview of the whole thoracic spine, but this miss small disk protrusions, and it remains invasive. Hence, MRI with sagittal imaging obviously is the first choice of imaging modality for investigation of the thoracic spine (Ross et al. 1987).

8.4.3 Thoracolumbar Junction

The thoracolumbar spine forms a junction between the thoracic spine, with limited mobility, and the lumbar spine, which is more mobile. Injuries to this level are frequent and are often serious. They are usually observed in high-speed traumata, including car-racing accidents, downhill skiing, and parachuting. Lesions may also result from falls such as in equestrian sports.

The lesions range from mild type 1 compression fractures to more serious type 4 fractures, with involvement of the posterior vertebral wall and often accompanied by disruption of the posterior ligamentous structure and subluxation or dislocation of the facet joints. Bone fragments in the spinal canal or disk protrusion may cause neural compression. Complete dislocation of the thoracolumbar junction can cause a complete disruption of the neural elements that is neurologically irreversible. The presence of bone fragments in the spinal canal is usually evaluated by CT scanning; MRI can be useful in the assessment of spinal cord compression and possible damage to the conus medullaris or cauda equina.

8.4.4 Lumbar Spine

The most frequently encountered osseous lesion to the lumbar spine is spondylolysis. Spondylolysis probably results from repeated stress around the pars interarticularis during hyperextension of the lumbar spine. This kind of movement is observed in golf or tennis serving, in downhill running, and in swimming (the butterfly breaststroke). Spondylolysis may be unilateral or bilateral and is usually diagnosed on standard X-ray examinations with lateral and oblique views; a defect in the neural arch should be sought that has been compared to a "scottie dog with a collar". Nevertheless, it may be extremely difficult to demonstrate unilateral spondylolysis, and bone scans may be useful for early detection of spondylolysis, especially in young athletes. Repeated trauma on unrecognized spondylolysis may lead to spondylolisthesis, with an anterior gliding of one vertebral body upon the other, possibly leading to radicular compression (Fig. 8.7).

Other osseous lesions to the lumbar spine again include compression fractures, ranging from type 1 to type 4. As for thoracic and cervical compression fractures, CT scanning is considered in type 4 fractures to outline the extent of the lesion and to determine the precise location of bony fragments in the spinal canal. Associated visceral injuries, such as contusion of the kidney, should always be sought in patients with severe compression fractures of the spine. Fractures of the transverse processes usually result from direct trauma sustained in football or in falls in equestrian sports. These lesions are well documented on plain roentgenograms. The role of MRI in the assessment of osseous lesions to the lumbar spine is less certain, although the examination may provide useful indications on possible compression lesions of the conus medullaris following extensive comminution fractures.

Although osseous lesions are observed in a wide range of sporting activities, the most common cause of consultation on the part of an amateur or professional sportsman is low back pain. The five lumbar vertebrae are massive, with short thick pedicles obviously designed to serve a weight-bearing function. Even more than at the cervical or thoracic levels the lumbar intervertebral disks act as shock absorbers. Disk disease remains a leading cause of chronic low back pain, affecting athletes and nonathletes alike and representing a major national health problem in many countries (Modic et al. 1983, 1984).

Medical imaging soon became involved in the objectivation of disk tissue tears, fissures, and possible herniation. The successive imaging techniques have greatly conributed to the understanding of morphological changes of the disks and to the mechanism of disk herniation (Athanosoulis and Lee 1987; Gelfand and Ott 1985).

The utility of plain roentgenograms in the diagnosis of lumbar disk lesions has become controversial since the appearance of more sophisticated techniques such as CT and MRI. Still, they remain a first-choice imaging modality in the evaluation of static disorders; they also reveal congenital or acquired lesions of the vertebral bodies and show osseous changes associated with severe disk degeneration. Advanced disk degeneration is suspected by the narrowing of the intervertebral disk space and is sometimes revealed by the presence of a linear lucency amidst the disk, known as the vacuum phenomenon.

Myelography was originally designed to delineate the spinal cord and the exiting nerve roots by opacifying the thecal sac with a contrast dys (now relatively safe, nonionic, and water soluble). A protruding disk causes an impingement on the opacified thecal sac; a possible compression of the exiting nerve root at the affected level can often be demonstrated as a widening of the this root as it emerges from the thecal sac. Nevertheless, the technique remains invasive and an indirect means for delineating the disk; it does not offer the possibility to evaluate the disk tissue itself. It often fails to demonstrate very lateral herniations of disk tissue in the neural foramen because of a lack of indentation on the contrast column in these cases (Haughton and Williams 1982; Hudgins 1970; Kieffer et al. 1969).

The advent of CT brought two major changes in the evaluation of disk pathology. It allowed direct visualization of the disk and measurement of its main density, and it allowed the acquisition of an axial image, thereby providing a much better approach of the disk, now visualized as a whole. The disk is now observed as a measurable difference between the higher density of disk tissue and the lower density of cerebrospinal fluid in the thecal sac, thus providing the necessary contrast for imaging. The disk has a homogeneous, relatively high density (on CT measurement, 50–80 HU); the spinal canal by contrast has a low density of fluid in the thecal sac (10–40 HU) and of epidural fat (density measurement between −50 and 0 HU). The posterior border of the disk evaluated on these axial images should be regularly and smoothly outlined; a disk herniation appears as a focal, irregular protrusion, angular or polypoid, of disk density into the spinal canal, sometimes associated with displacement or thickening of the dorsal root ganglion or the exiting nerve root (Haughton and Williams 1982).

CT, however, does not measure any density difference between the central part of the disk and its outer layers. CT thus cannot differentiate the inner nucleus pulposus from the surrounding annulus fibrosus and therefore remains relatively insensitive to early degenerative changes as long as the disk morphology has not changed. This may result in underestimation of the severity of such degenerative changes. On the other hand, CT enables precise evaluation of the sequelae of disk degeneration, with a clear dipiction of facet joint disease, ligamentous hypertrophy, and stenosis of the spinal canal. Currently, CT remains the primary method of screening for degenerative disk disease since it is noninvasive, cost-effective, and offers an overall accuracy of 80%–90% in the diagnosis of disk herniation. The major disadvantages of the technique are the limited field of view due to axial scanning and the still limited contrast sensitivity between the various soft tissues (Modic et al. 1986b).

Discography is a technique in which a small amount of contrast dye (1.5 ml) is injected into the disk. This injection is often performed without difficulty when there is a certain degree of disk degeneration and hence less resistance of the loosened fibers of the annulus fibrosus. Extravasation of contrast medium occurs in the fissures and tears of the degenerated disk. In some cases, however, the method cannot differentiate clearly between diffuse disk degeneration and disk herniation. A better result is obtained when the method is combined with subsequent axial CT scan. Still, the method is traumatic and has justifiably fallen out of favor (Kieffer et al. 1969; Modic et al. 1984).

The Disk Anatomy. The intervertebral disk is composed of three distinct parts: the inner nucleus pulposus, the surrounding annulus fibrosus, and the cartilaginous endplate. The nucleus pulposus represents the remnants of the embryonal notochord; this consists of well-hydrated, mucoid, gelatinous material derived from degenerated notochord cells and this fibrocartilaginous strands. This essentially incompressible gelatinous structure redirects forces from a vertical plane to a horizontal one. The nucleus pulposus is thereby the major component of the load-bearing capacity of the lumbar intervertebral disk. Peripherally the nucleus bends imperceptibly with the annulus fibrosus without clear demarcation between them. Encircling the nucleus, the annulus fibrosus serves as a limiting capsule of the nucleus. It consists of several layers of tough collagenous fibers, more closely packed anteriorly than posteriorly and thus providing greater strength anteriorly. The thinner, more vertically oriented fibers of the posterolateral parts of the disk are weaker; disk herniations, indeed, more often occur posterolaterally. The annulus is completely circular and attaches to the adjacent vertebral bodies superiorly and

a

b

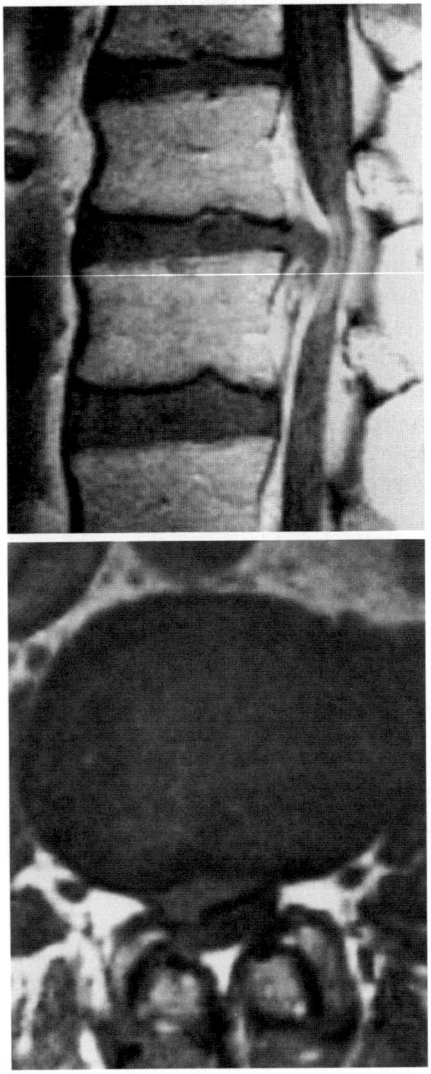

Fig. 8.8 a, b. Lumbar spine, surface coil. **a** Sagittal T_1-weighted image (SE 600/15, 3 mm), **b** Axial T_1-weighted image (SE 600/15,4 mm). Disk herniation L2–L3; the axial images localize the posterolateral herniation and clearly show the impingement of the herniated disk on the epidural fat and the thecal sac

inferiorly by the Sharpey's fibers at the site of the fused epiphyseal ring. Different types of collagen fibers are found in the annulus, providing the necessary resistance to radial tension induced by axial loading on the weight-bearing disk and translated by the nucleus pulposus. The superior and inferior surfaces of both annulus and nucleus are covered by a thin plate of hyaline cartilage. While this cartilaginous endplate contains several blood vessels at birth, it is completely avascular in the adult human body although it remains metabolically active and a major site of diffusion with the spongiosa of the adjacent vertebral bodies.

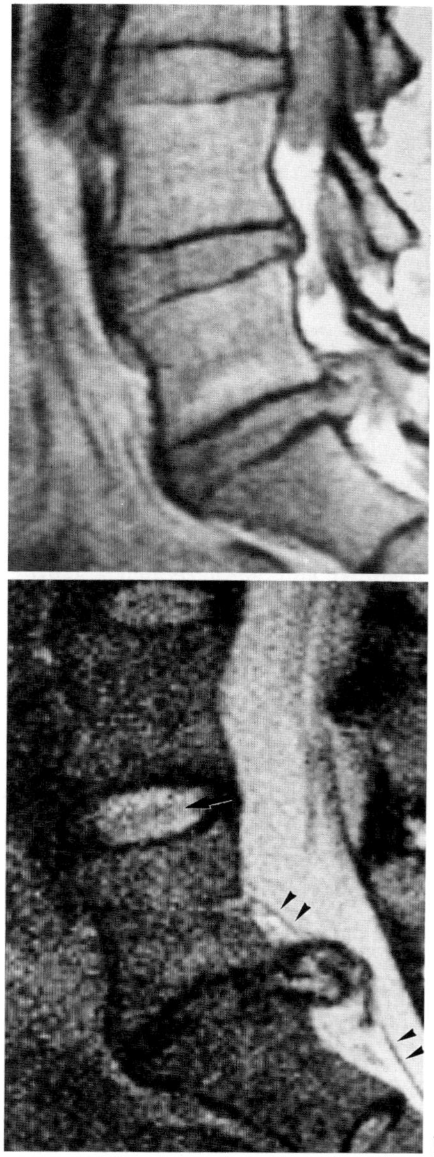

a

b

Fig. 8.9 a, b. Lumbar spine, surface coil. **a** Midsagittal T_1-weighted image (SE 600/15). **b** Same plane, T_2-weighted image (SE 2200/90). Extruded disk herniation L5–S1); note the deformity of the thecal sac (*arrow*). Signal loss of the herniated disk on the T_2-weighted image; normal appearance of the disks L3–L4 and L4–L5 with a typical internuclear cleft (*arrowheads*)

Degeneration of the disk starts early in infancy, with biochemical and morphological changes. These degenerative changes can be divided roughly into two basic processes. The first is characterized by the breaking of the collagen fibers of the annulus fibrosus. In the normal juvenile spine the annulus fibers are even more resistant to trauma than the vertebral bodies; however, by the second decade of life trauma may selectively rupture these annulus fibers.

As they are weaker in the posterolateral parts of the disk, they usually break at the points first. The nucleus then migrates through the different disrupted layers of the annulus, usually in posterolateral direction. When all annulus layers are disrupted, the nucleus herniates outside the disk to a position beneath the posterior longitudinal ligament (subligamentous disk herniation) or may even break through the ligament and may, in rare cases, even penetrate the dural sac (intradural disk herniation). The second degenerative process consists merely of biochemical changes of the mucoid material of the nucleus pulposus. It loses mucopolysaccharides, protein, and especially water, causing it to shrink and to lose its elasticity. As the nucleus shrinks, the fibers of the surrounding annulus become more slack and tend to buckle into the nucleus while the peripheral layers of the annulus bulge outward, which is described as a global protrusion of the disk or a "bulging annulus."

Disk herniations or protrusions can easily be documented on T_1-weighted images, sagittal or axial; here their morphological changes are obvious (Fig. 8.8). An impingement on the exiting nerve root is visualized at the level of the neural foramen. A T_1-weighted spin-echo sequence, however, does not show any difference between central and peripheral disk tissue. Therefore a T_2-weighted spin-echo sequence with long repetition time and echo time is necessary, this being more sensitive to slight differences in hydrogen content of the different tissues. The nucleus pulposus, having a higher water content than the surrounding annulus should logically present a higher signal intensity than the surrounding annulus. The disk, however, appears on a T_2-weighted spin-echo sequence as a biconvex structure with high signal intensity and a central, thin, horizontal band with lower signal intensity (Fig. 8.9). In fact, this is a normal finding in all patients aged 30 or over. It is called the internuclear cleft and is thought to represent the invagination of loosened fibers of the annulus into the shrinking nucleus (Modic et al. 1984).

Progressive disk degeneration is still best documented on these T_2-weighted spin-echo images. A degenerated disk indeed loses most of its water content and has a characteristic low-intensity signal, contrasting with the bright signal of a normal, well-hydrated intervertebral disk (Fig. 8.9b).

8.5 Future Prospects

The greatest advantages of MRI over CT are its excellent contrast resolution, its multiplanar imaging capacity, and a total lack of ionizing radiation. The combination of axial images (as in CT) and sagittal images (as in myelography) makes MRI the ideal means for evaluating the spine.

There are, however, some major drawbacks related to the relatively longer examination times, the high cost of MRI installations, their occasional lack of availability, and the confined space in the magnet. Critically ill patients and claustrophobic patients cannot be examined by MRI. Metallic material such as Harrington rods can lead to severe artifacts, making MRI impossible.

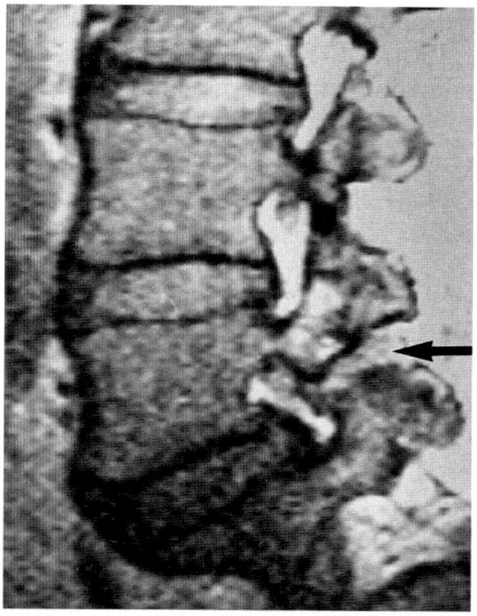

Fig. 8.10. Lumbar spine, surface coil. Parasagittal T_1-weighted spin-echo image (SE 600/15). Spondylolysis with an obvious defect in the pars interarticularis of the neural arch of L 5 (*arrow*). Note the slippage of L 5 and the resulting deformity of the intervertebral foramen as well as the degenerative signal loss of L 5–S 1

Although the available space in the magnet is not likely to be improved, the acquisition times and the global examination time will certainly be reduced significantly by the use of new sequences, especially the so-called gradient-echo sequences now currently under investigation (Fig. 8.10). Up to now spin-echo sequences have been applied for most clinical indications, including those in the spine. These spin-echo sequences indeed have good signal-to-noise and contrast-to-noise ratios, providing high-quality images, but they still require relatively long examination times. The gradient-echo sequences or gradient-recalled echo sequences are much faster; in fact, the major difference lies in the initial radio frequency pulse applied to the tissue. For spin-echo sequences a 90° pulse is applied, for gradient-echo imaging a pulse between 0° and 90° (called a flip angle). If, for instance, a flip angle of 20° is used, the net magnetization vector is tipped only 20° from the direction of the externally applied magnetic field; most of the net magnetization thus remains longitudinal. This is an efficient way of producing an MR signal with a relatively short repetition time, thereby shortening the acquisition time significantly. T_1- and T_2-weighting effects can be obtained by varying the flip angles (Perkins and Wehrli 1986; Stark et al. 1987).

Other technical improvements include a peripheral pulse gating instead of an electrographic gating. Cardiac gating is, as mentioned above, indispensible in the investigation of cervical and thoracic spine to reduce motion related to cerebrospinal fluid and vascular pulsations. Up to now, an electroradiographic gating has been used which triggers the gating to some phase of the cardiac cycle. Electrodes must thus be placed on the patient's chest, which is a time-consuming procedure. A simpler way consists in attaching a pulse photosensor to a digit and triggering the gating to either systole or diastole.

Contrast agents for MRI examinations are now available. Gadolinium DTPA or DTOA is a paramagnetic contrast dye that shortens the T_1 and T_2 relaxation times significantly; it is particularly useful in the evaluation of spinal tumors and discitis but may also prove useful in differentiating postoperative scarring from recurrence of disk herniation.

Lastly, three-dimensional imaging procedures are being developed. It seems likely that three-dimensional data acquisition with thin slices of 1–1.5 mm and secondary multiplanar reconstructions will replace the present time-consuming procedures in the very near future. These procedures provide superior detail resolution, and as the reconstructions can be obtained in all spatial planes, lesions involving the ligaments will probably be diagnosed more efficiently by MRI.

These technical improvements together with increasing demands for noninvasive imaging, if possible on an outpatient basis, will allow MRI to replace myelography and at least to compete with CT, if not to replace it altogether. As the resolution is improving and the diagnosis of smaller lesions in an earlier phase becomes possible, MRI is likely to become a major tool in the diagnosis of sport-related lesions but also in the prevention of these lesions.

References

Athanosoulis CA, Lee AK (1987) Algorithms in radiology and medical imaging. Radiology 164:565–567

Bale JF Jr, Bell WE, Dunn V et al. (1986) Magnetic resonance imaging of the spine in children. Arch Neurol 43:1253–1256

Brant-Zawadzki M, Norman D (1987) Magnetic resonance imaging of the central nervous system. Raven, New York, pp 289–328

Brodsky AE, Binder WF (1979) Lumbar discography: its value in diagnosis and treatment of lumbar disk lesions. Spine 4:110–120

Burnett KR, Levine JB (1986) MR of lumbar disk disease. Appl Radiol 15:133–141

Chafetz NI, Genant HK, Moon KL et al. (1984) Recognition of lumbar disk herniation with NMR. AJNR 5:23–26

Chakeres DW, Fliekinger F, Bresnahan JC et al. (1987) MR imaging of acute spinal cord trauma. AJNR 8:5–10

Czervionke LF, Daniels DL (1988) Cervical anatomy and pathologic processes. Applications of new MR imaging techniques. Radiol Clin North Am 26 (5)

Daniels DL, Grogan JP, Johansen JG, Meyer GA, Williams AL, Haughton VM (1984) Cervical radiculopathy; computed tomography and myelography compared. Radiology 151:109–113

Daniels DL, Hyde JS, Kneeland JB et al. (1986) The cervical nerves and foramina: local-coil imaging. AJNR 7:129–133

Daniels DL, Haughton VM, Naidich (1987) Cranial and spinal magnetic resonance imaging: an atlas and guide. Raven, New York

Edelman RR, Shoukimas GM, Stark MD et al. (1985) High resolution surface coil imaging of the lumbar disk disease. AJNR 6:479–485

Eisenberg RL, Amberg JA (1981) Critical diagnostic pathways in radiology: an algorythmic approach. Lippincott, Philadelphia, pp 263–279

Firooznia H, Benjamin V, Kricheff II, Columbu C (1984) CT of lumbar disk herniation: correlation with surgical findings. AJR 142:587–592

Franken EA Jr, Berbaum KS, Dunn V et al. (1986) Impact of MR imaging on clinical diagnosis and management: a prospective study. Radiology 161:377–380

Gelfand DW, Ott DJ (1985) Methodologic considerations in comparing imaging methods. AJR 144:1117–1121

Grenier N, Grossman RI, Scheibler ML et al. (1987) Degenerative lumbar disc disease: pitfalls and usefulness of MR imaging in detection of vacuum phenomena. Radiology 164:861–865

Han JS, Kaufman B, EL Yousef SJ et al. (1983) NMR imaging of the spine. AJR 141:1136–1145

Haughton VM, Williams AL (1982) Computed tomography of the spine. Mosby, St Louis

Hudgins WR (1970) The predictive value of myelography in the diagnosis of ruptured lumbar disks. J Neurosurg 32:152–162

Kieffer SA, Stadlan EM, Mohandas A et al. (1969) Diskographic-anatomical correlation of developmental changes with age in the intervertebral disc. Acta Radiol Diagn 9:733–739

Kulkarmi MV et al. (1985) Diagnosis of spinal arteriovenous malformation in a pregnant patient by MR imaging. J Comput Assist Tomogr 9:171–173

Lipson SJ, Muir H (1981) Experimental intervertebral disk degeneration. Spine 6:194–210

Masaryk TJ, Ross JS, Modic MT et al. (1988) High resolution MR imaging of sequestered lumbar intervertebral disks. AJNR 9:351–358

McArdle CB, Crofford MJ, Mirfakhrau M (1986) Surface coil MR of spinal trauma: preliminary experience. AJNR 7:885–893

McCunnan BL (1987) Low osmolarity contrast media: premises and promises. Radiology 162:1–8

Mills TC, Ortendahl DA, Hylton NM et al. (1987) Partial flip angle MR imaging. Radiology 162:531–539

Modic MT, Weinstein MA, Pavlicek W et al. (1983) Magnetic resonance imaging of the cervical spine. Technical and clinical observations. AJR 141:1129–1136

Modic MT, Weinstein MA, Pavlicek W et al. (1983) Nuclear magnetic resonance imaging of the spine. Radiology 148:757–762

Modic MT, Pavlicek W, Weinstein MA et al. (1984) Magnetic resonance imaging of intervertebral disk disease. Radiology 152:103–111

Modic MT, Masaryk TJ, Mulopulos GP et al. (1986a) Cervical myelopathy: prospective evaluation with surface coil imaging, CT with metrizamiede and metrizamiede myelography. Radiology 161:753–754

Modic MT, Masaryk TJ, Paushter D (1986b) Magnetic resonance imaging of the spine. Radiol Clin North Am 14:229–245

Pech PE, Haughton VM (1985) Lumbar intervertebral disk: correlated MR and anatomic study. Radiology 156:699–701

Perkins TG, Wehrli FW (1986) CSF signal enhancement in short TR gradient echo images. Magn Reson Imaging 4:465–467

Quencer RM et al. (1986) Magnetic resonance imaging of the chronically injured cervical spinal cord. AJNR 7:457–464

Resnick D (1985) Degenerative diseases of the vertebral column. Radiology 156:3–14

Ross JS, Perz-Reyes N, Masaryk TJ et al. (1987) Herniated thoracic intervertebral disks: assessment by MR. Radiology 165:511–513

Runge VN, Clanton JA, Lukehart CM et al. (1983) Paramagnetic agents for contrast-enhanced MR: a review. AJR 141:1209–1215

Stark DD, Bradley WG (1988) Magnetic resonance imaging. Mosby, St Louis

Stark DD, Hendrick RE, Hahn PF et al. (1987) Motion artefact reduction with fast spin-echo imaging. Radiology 164: 183–191

Steiner RE (1985) Magnetic resonance imaging: its impact on diagnostic radiology. AJR 145:883–893

Tarr RW, Drolshagen LF, Kerner TC et al. (1987) MR imaging of recent spinal trauma. J Comput Assist Tomogr 11:412–417

Teresi LM, Lufkin RB, Reicher MA et al. (1987) Asymptomatic degenerative disk disease and spondylosis of the cervical spine: MR imaging. Radiology 164:83–88

Wehrli FW, MacFall JR, Newton TH (1983) Parameters determining the appearance of MR images. In: Newton TH, Potts DG (eds) Modern neuroradiology, vol 2. Clavadell, San Anselmo

Williams AL (1983) CT diagnosis of degenerative disc disease. Radiol Clin North Am 21:289–300

Trends

9 MR Perspectives in Sports Cardiology

Y. Taeymans and R. Luypaert

9.1 Introduction

Regular exercise is associated with a variety of physiological adaptations, of which the most apparent is bradycardia at rest due to increased vagal tone. Increased aerobic capacity, increase in left ventricular volume, and ventricular wall thickness are the other most frequently cited adaptations to exercise. Whether volume or wall thickness adaptation is more important seems to depend on the type of exercise. Indeed, noninvasive studies have shown regular dynamic exercise to be associated with volume overload and with maintenance of the usual ratio of wall thickness to internal diameter (Roeske et al. 1975; Peronnet et al. 1981). On the other hand, more pronounced changes in wall thickness were observed with static exercise (Morganroth et al. 1975). Actually, it is not known what can be considered a safe hypertrophy. Attained at an early age, it seems to be long-lasting (Saltin and Grimby 1968). All these studies were carried out using echocardiography. This is indeed a simple and noninvasive method, but it lacks precision concerning quantification. The volume of the left ventricle is estimated from a single cubed linear dimension (Feigenbaum et al. 1972) usually corrected for the effect of the size (Teicholz et al. 1976). A more accurate method using Simpson's rule was developed later (Schiller et al. 1979). Still, left ventricular stroke volume calculations remain very difficult because of the limited number of acoustic windows. Right ventricular stroke volume calculations are even less reliable due to incomplete visualization of the right ventricle.

In sports cardiology four main parameters for both left and right ventricles are needed: precise ventricular volume, precise ventricular mass, contractility, and relaxation. The two last dynamic parameters are especially useful in the case of abnormal volume or wall thickness. None of these parameters, however, is routinely available. Cardiac magnetic resonance imaging (MRI) has potentially major advantages compared to other imaging techniques. It may bring the clinical data needed to study accurately and longitudinally the effects of dynamic exercise on cardiac function. It is noninvasive and produces clearly defined anatomic images of the heart with a sharp contrast between blood, myocardium, pericardium, and epicardial fat. MRI is not limited regarding imaging planes or access windows. Indeed, axial, frontal sagittal, oblique, and even double-oblique imaging planes can be obtained. Unlike in biplane echocardiography, the complete heart and great vessels are seen

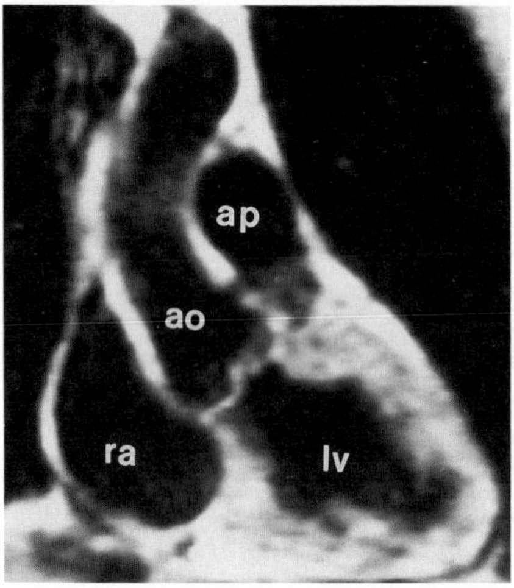

Fig. 9.1. Coronal view of a normal heart, with good visualization of the left ventricular cavity (*lv*), right atrium (*ra*), aorta ascendens (*ao*), and pulmonary artery (*ap*)

(Fig. 9.1). Obesity is not a problem as long as the patient can be installed confortably in the imager. The only limitation is the cardiac rhythm: no imaging is possible in the case of severe arrhythmia. Cost and time needed to obtain images are nontechnical limitations.

9.2 Cardiac Imaging Techniques

As explained in other chapters of this volume, an MR image is not obtained instantaneously but is the result of the acquisition of, typically, 256 or 128 lines of 256 pixels. Since the heart and blood are in almost constant movement, a synchronized acquisition is mandatory. This synchronization is achieved by detecting the R wave of an ECG. Special equipment is needed to be able to record the ECG in a high magnetic field with rapidly switching gradients and radio frequency signals. This requires proper filtering, disconnection synchronized with gradient switching, optical insulation, and nonmetallic electrodes. Each of the 128 lines is acquired during the same phase of the cardiac cycle and usually averaged four times. The consequence is a prolongation of the acquisition time. A spin-echo sequence with TE of 22 ms requires an acquisition time of 4–8 min, depending mainly on the duration of the cardiac cycle. One is, however, not limited to one slice during the cardiac cycle. Multislice acquisition is possible, each slice with the same TE and TR (= cardiac cycle duration) but at a different phase of the cardiac cycle. Real three-dimensional imaging (all images at the same phase of the cardiac cycle) is actually not possible with a standard technique due to an extremely prolonged acquisition time (several hours). This fundamental physical limit is not really important as far as diagnostic imaging is concerned. Using a fast-acquisition

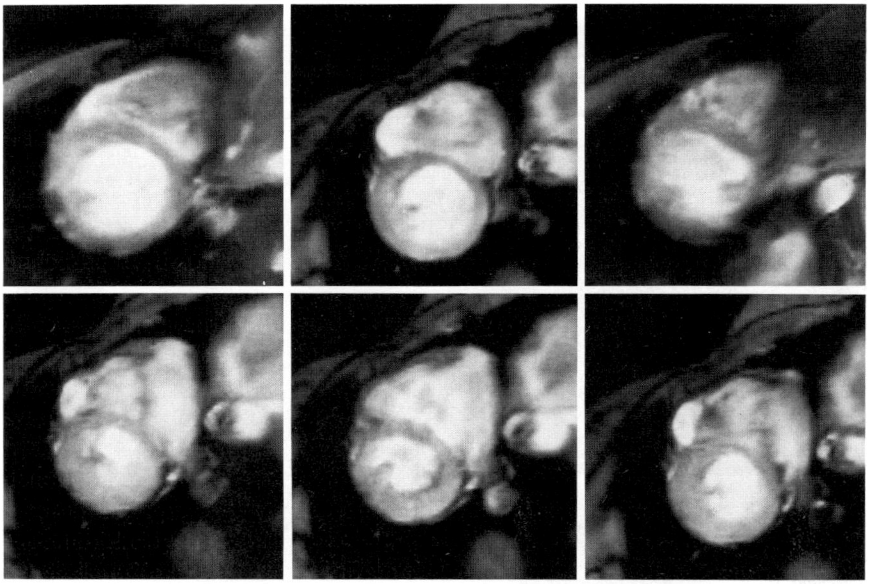

Fig. 9.2. Short axis multiphase images of the left ventricle at the level of the papilary muscles

technique with a low flip angle (FLASH) and TE of 12 ms, up to 25 images can be obtained during the cardiac cycle at the same slice position. Projected in a closed cine loop, dynamic images are thus obtained at a particular slice position.

9.3 Quantification of Ventricular Volume and Mass

The myocardium is clearly delineated using the routine diagnostic imaging technique (multislice spine-echo), allowing contour detection. Theoretically this is not suitable for volume calculations. Using this acquisition technique may result in one or two slices in diastole and others in systole, each at a different level. Simpson's rule calculation, integrating the left ventricular surface of each slice, might take into account some portions of the left ventricle twice and others not. In practice, surprisingly, this does not seem to matter very much (Shapiro et al. 1989), and the total acquisition time is markedly shorter as compared to diastolic images only. A second approach requires a multiphase multislice acquisition technique (FLASH). Up to 25 images can be obtained divided in three or in four slices with eight or six phases, respectively (Fig. 9.2). Increasing the number of slices tends to produce poorly contrasted images. This sequence must therefore be repeated with a slice shift to be able cover the entire heart. Unlike in spin-echo images, blood appears very intense, allowing a good delineation of the ventricular cavities. Both left ventricular and

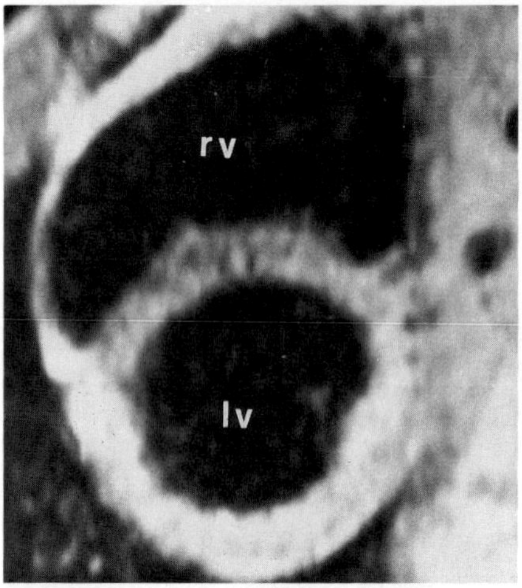

Fig. 9.3. Short axis view of left (*lv*) and right ventricle (*rv*)

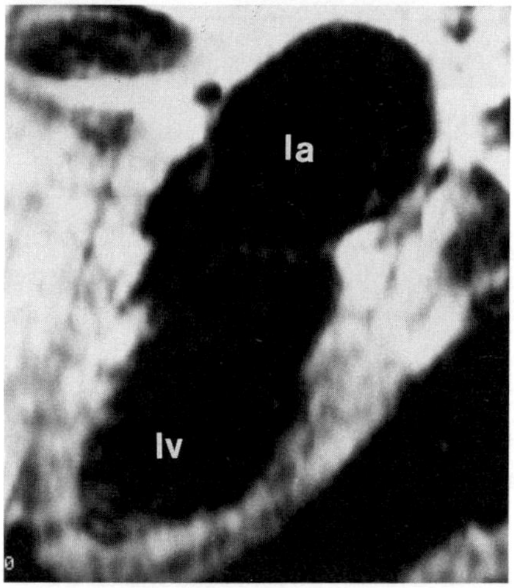

Fig. 9.4. Two chamber view of a normal heart, with visualization of the left ventricle (*lv*) and left atrium (*la*)

right ventricular volumes can be measured accurately in a reasonable total imaging time (Markiewics et al. 1987; Sechtem et al. 1987a, b). The high correlation between right and left stroke volumes suggests this method to be reliable. The main advantage of this technique is the measurement of contractility and relaxation parameters based on the evolution of real volumes.

Standard MRI planes are oblique to the different cardiac structures. Oblique planes equivalent to those defined for echocardiography are mandatory only if one is interested in the dimensions of the ventricles or the wall thickness. A short axis view corresponds to a plane perpendicular to the long axis of the heart (Fig. 9.3). A two-chamber view corresponds to a plane sagittal to the septum (Fig. 9.4), and a four-chamber view corresponds to a plane perpendicular to the septum (Fig. 9.5). Unlike echocardiography, which

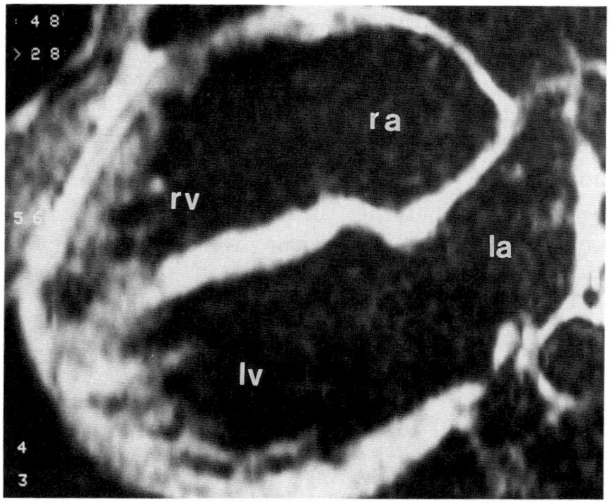

Fig. 9.5. Four-chamber view of the heart, with visualization of left ventricle (*lv*), left atrium (*la*), right ventricle (*rv*) and right atrium (*ra*)

permits orientation by direct visualization, MR needs one or more "scout" views to be able to derive the spatial orientation of the septum and generate the pulse sequences needed to obtain oblique planes.

Initially a long axis series was made by rotating at a fixed 30° from the coronal plane (as apposed to a 30° right anterior oblique angiographic view). Short axis planes were obtained perpendicular to this first plane (Dinsmore et al. 1984). To limit total imaging time a methodology was developed to be able to derive the three standard echocardiographic planes, adapted to each individual patient, from an axial multislice acquisition sequence (Luypaert et al. 1988). In a total imaging time of about 45 min, one scout view and three spin-echo multislice oblique image series along the three major orientations can be obtained. Dimensions of left and right ventricle and left and right atrium can be measured and correlated to those known from echocardiography (Weyman 1982).

Thus, at the present stage of technology, some of the tools needed in sport cardiology are now available. However, due to the rather limited availability of this technique no extensive experience has been gained so far in the specific field of sports cardiology. Left ventricular hypertrophy, idiopathic or hypertension related, has been studied and quantified to some extent by MRI (Eichstaedt et al. 1989). MR images have been shown to be superior especially in patients with hypertrophic cardiomyopathy. The hypertrophy is, indeed, usually very variable in localization and extent (Higgins et al. 1984). Only visualization of the whole left ventricle as obtained by MRI allows a correct diagnosis.

9.4 Conclusion

Although not accepted as a standard technique, MRI of the heart may represent the ideal method in the future regarding quantification of both ventricular volumes and masses and that of ventricular dynamics. Essentially noninvasive, it may be an important tool in sports medicine. At the present stage of the technique, all the techniques needed are available, but they have not yet proven their clinical utility. Future developments are, however, expected to make it more cost effective in terms of investment and time.

References

Dinsmore RE, Wismer GL, Levine RA et al. (1984) Magnetic resonance imaging of the heart: positioning and gradient angle selection for optimal imaging planes. AJR 143:1135–1142

Eichstaedt H, Danne O, Langer M, Cordes M, Schubert C, Felix R, Schmutzler H (1989) Regression of left ventricular hypertrophy under ramipril treatment investigated by nuclear magnetic resonance imaging. J Cardiovasc Pharmacol V 13 (3):75–80

Feigenbaum H, Popp RL, Wolfe SB (1972) Ultrasound measurements of the left ventricle: a correlative study with angiography. Arch Intern Med 129:461

Higgins CB, Byrd BF, McNamara MT et al. (1984) Magnetic resonance imaging of hypertrophic cardiomyopathy. Circulation 70 [Suppl 2]:248

Luypaert R, Taeymans Y, van Cauteren M, Verhelle F, Peeters F, Osteaux M (1988) Magnetic resonance imaging of double oblique slices through the human heart. SPIE Biostereometrics 1030:237–244

Markiewics W, Sechtem U, Kirby R, Derugin N, Caputo GC, Higgins CB (1987) Measurements of ventricular volumes in the dog by nuclear magnetic resonance imaging. J Am Coll Cardiol 10 (1):170–177

Morganroth J, Maron GJ, Henry WL et al. (1975) Comparative left ventricular dimensions in trained athletes. Ann Intern Med 82:521–524

Peronnet F, Ferguson RJ, Perrault H et al. (1981) Echocardiography and the athlete's heart. Phys Sports Med 9:102–112

Roeske WR, O'Rouke RA, Klein A et al. (1975) Noninvasive evaluation of ventricular hypertrophy in professional athletes. Circulation 53:286–292

Rushmer RF, (1970) Cardiovascular dynamics. Saunders, Philadelphia, pp 70–112

Saltin B, Grimby G (1968) Physiological analysis of middle aged and old former athletes: comparison with still active athletes of the same age. Circulation 381:104–115

Schiller N et al. (1979) Left ventricular volume from paired biplane two-dimensional echocardiography. Circulation 60:547

Sechtem U, Pflugfelder PW, White RD, Gould RG, Holt W, Lipton MJ, Higgins CB (1987 b) Cine MRI: potential for the evaluation of cardiovascular function. AJR 148:239–246

Sechtem U, Pflugfelder PW, Cassidy MM, White RD, Cheitlin MD, Schiller NB, Higgins CB (1988) Mitral or aortic regurgitation: quantification of regurgitant volumes with cine MR imaging. Radiology 167 (2):425–455

Shapiro RP, Rogers WJ, Beyar R, Soulen RL, Zerhoumi EA, Lima JAC, Weiss JL (1989) Determination of left ventricular mass by magnetic resonance imaging in hearts deformed by acute infarction. Circulation 79:706–711

Stratemeier EJ, Thompson R, Brady TJ, Miller SW, Saini S, Wismer GL, Okada RD, Dinsmore RE (1986) Ejection fraction determination by MR imaging: comparison with left ventricular angiography. Radiology 158:775–777

Teicholz LE, Keulen T, Herman MV, Gorlin R (1976) Problems in echocardiographic volume determinations: echocardiographic-angiographic correlations in the presence or absence of asynergy. Am J Cardiol 37:7

Weyman AE (1982) Cross-sectional echocardiography. Lea and Febiger, Philadelphia

10 ^{31}P Magnetic Resonance Spectroscopy of Muscle: The Missing Link Between Physiology and Sports Practice

E. ACHTEN, K. VANDENBORNE, M. OSTEAUX, and K. DE MEIRLEIR

10.1 Introduction

Muscle is the motor of movement. The performance of an individual is highly dependent upon the adequate output of force by his muscles. In the world of sports physiology muscle is the system with the fastest changes in performance because it can be trained in a few weeks to become several times as powerful as initially. However, if a sufficient level of exercise is not maintained, it quickly loses its force. Force development is not the only adaptation in muscle related to exercise; the efficiency of oxygen consumption, adaptation to the use of more efficient fuel sources, changes related to different muscle fibers, and innervation play key roles in the performance of athletes.

The causative factors in sports injuries are often related to muscle problems. Excessive activity leads to local tissue breakdown, resulting in fatigue and overload. Fatigue, previous injuries, cold, viral infections, overuse, and insufficient training results in muscle soreness, weakness, or imbalance, increasing the risk of intrinsic trauma to muscle, tendon, ligaments, or joint.

The ever greater need for extreme performance and for beating records has introduced numerous performance boosters in the area of sports. Apart from improvements in physical and psychological aids, all other ergogenic aids (Grisogono and Yaffé 1986) increase work output or performance by direct action upon muscle. It is therefore not surprising that substantial research efforts have been made to increase the performance of muscle by action upon the biochemical events responsible for force development or for a more efficient energy supply. To test the influence of various factors on this performance it is essential to have a good probe for the biochemical events involved. The acquisition of biochemical information from muscle was possible until recently only by means of a painful muscle biopsy. This invasive tool was the only link between the fundamental biochemistry of the laboratory and the working muscle of the athlete. Moreover, this allows only a snapshot impression of the metabolic events involved in the energy supply of the working muscle; due to its invasiveness it can not be applied for sequential follow-up.

Magnetic resonance spectroscopy (MRS) is a biochemical sampling technique that permits the repetitive, continuous, and noninvasive acquisition of metabolic changes in the working muscle. It is, in fact, the missing link between sports physiology and sports practice.

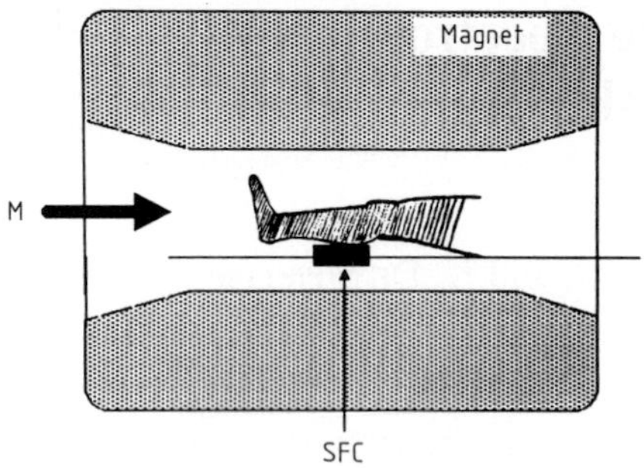

Fig. 10.1. Schematic drawing of the position of the subject in the magnet bore during MRS of the calf muscles. *M*, Magnetic Field. Excitation of the spins and reception of the emitted MR signals is done by the surface coil (*SFC*) under the calf muscles

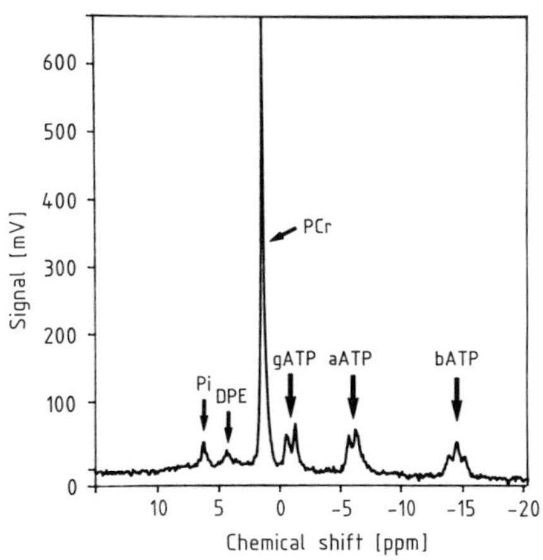

Fig. 10.2. Phosphorous MR spectrum of human calf muscles in rest. Six peaks can easily be seen, from left to right: inorganic phosphorous (*Pi*), diphosphoesters (*DPE*), phosphocreatine (*PCr*), gamma-, alpha-, and beta-ATP. The area under the peak is a measure for the relative concentration of the corresponding metabolites. At rest, the concentration of sugar phosphates is low and undetectable with MRS

10.2 Phosphorous Magnetic Resonance Spectroscopy of Skeletal Muscle

In order to extract biochemical information with MRS, the muscle to be studied, in this case the calf muscle group, is placed in the center of a magnetic field (Fig. 10.1). Under the muscles we place an antenna or surface coil in order to able to excite the spins (e.g. the phosphorous atoms present in certain chemical species such as ATP) locally in the muscle and to collect the signal that they emit when they return to equilibrium (see Chap. 1). Figure 10.2 shows the ^{31}P spectrum from the calf muscles at rest. In such a spectrum, the x-axis represents the relative frequency and the y-axis the magnitude of the MRS signal. Different chemical species (e.g., PCr, ATP) resonate at different frequencies and have distinct positions on the x-axis. The reference frequency is often chosen to be the resonant frequency of PCr, and the relative frequency difference of the other resonances with respect to PCr is then expressed in parts per million (ppm).

With a sufficient number of accumulations, the spectrum of the resting muscle has five major peaks: inorganic phosphate (Pi), phosphocreatine (PCr), and the three adenosine triphosphate (ATP) peaks – gamma, alpha, and beta. In some cases two other peaks are present: a phosphodiester peak, which is invariable during the exercise and represents membrane metabolites, and a phosphomonoester peak, which indicates anaerobic glycolysis. It is important to remember that MRS can detect only metabolites that are freely in solution or those in rapid exchange with the free pool (Gadian 1982). This means that ^{31}P MRS spectra show no ADP because most ADP is bound. Furthermore, the free concentration of ADP is less than 10 μmol/l at rest, and to be detected with MRS the concentration of a ^{31}P-containing metabolite should be higher than 1 mmol/l.

Several biochemical properties of the energy metabolites can be deduced from a spectrum. The area under a given peak is proportional to the free concentration of the corresponding metabolite. Another parameter that can easily be assessed with ^{31}P MRS is the intracellular pH of the muscle (Petroff et al. 1985). This is very convenient and important because intracellular pH in muscle is a good barometer for anaerobic glycolysis. Figure 10.3 shows what happens if we change the pH of a phantom solution containing PCr and Pi. The PCr peak remains at the same frequency, but when we titrate the solution with HCl, the pH declines, and the Pi peak gradually shifts toward the PCr peak. Figure 10.4 shows the relation between the pH of the solution and the chemical shift of Pi versus PCr. The curve corresponds to the Henderson-Hasselbach equation for the equilibrium reaction between HPO_4^{2-} and $H_2PO_4^-$. The pK_a of this acid-base couple is 6.8. This means that the biggest changes in relative concentration are in the physiological pH range. Because of a rapid exchange between the two Pi species, their MRS signals merge together as a single line in a position depending on their relative concentrations, or in other words, depending on the pH of the solution. This property allows the noninvasive measurement of intracellular pH in muscle and many other organs.

Due to the imprecise definition of the volume where the MRS signals originate, it is difficult to measure absolute concentrations. Because we want to see changes over time rather than static information, relative concentration changes or changes in certain ratios are far more important and are easily assessed with sequential MRS.

Rest spectra in 15 normal healthy subjects showed a mean relative ATP content of 19% (SD = 4%) of the total MRS-visible ^{31}P, a mean relative PCr content of 68% (SD = 5%), and a mean relative Pi content of 10%, the rest being phosphodiesters and phosphomonoesters. The absolute concentrations in vivo are sometimes estimated starting from the intracellular ATP concentration, which is relatively constant at 8.5 mmol/l (Taylor et al. 1986; Dawson 1982). With this assumption, the mean concentrations for PCr and Pi in the calf muscles of our 15 volunteers were approximately 30 mmol/l and 4.5 mmol/l at rest, respectively.

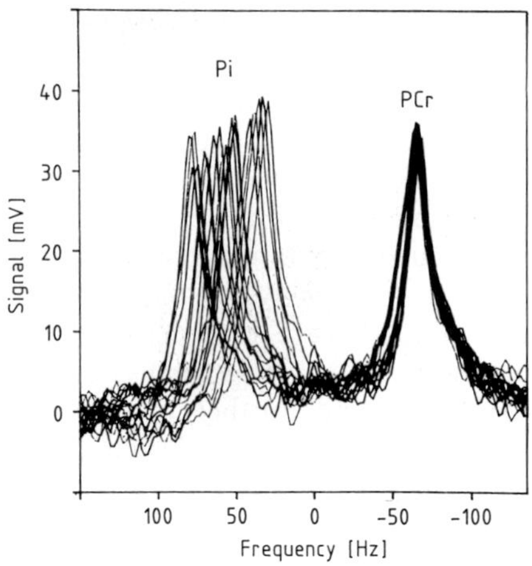

Fig. 10.3. Superimposition of spectra acquired from a phantom containing PCr and Pi. When the pH of the solution is changed, the resonance frequency of PCr remains the same while the resonance frequency of Pi lowers when the solution becomes more acidic

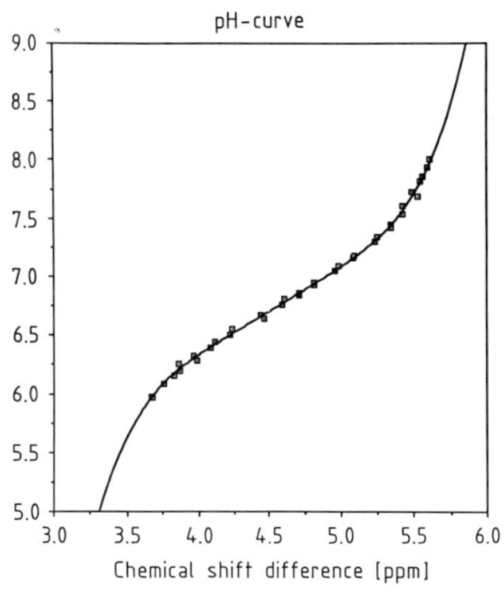

Fig. 10.4. pH titration curve from the experiment in Fig. 10.4. The x-axis displays the distance between PCr and Pi expressed in parts per million of the resonance frequency of PCr. The y-axis is the pH of the solution measured with a pH-meter. From this, the pH in the muscle cells can be calculated using the following form of the Henderson-Hasselbach equation: pH 6.77 log [(Pi shift -3.29)/(5.68$-$Pi shift)], derived from our pH experiment

10.3 Dynamic ^{31}P MRS of Muscle

Rest spectra are seldom abnormal. Only in patients with pain at rest due to obstructive arterial disease (Radda 1988), patients with mitochondrial (Arnold et al. 1985) and dystrophic myopathies (Newman et al. 1982), and subjects with muscle injury after excessive exercise (McCully et al. 1988c) can a rest spectrum show some abnormality. It is far more interesting to study the dynamic changes in the relative concentrations of the metabolites and the pH during and after exercise. For this purpose, the subject or patient is positioned supine, with the

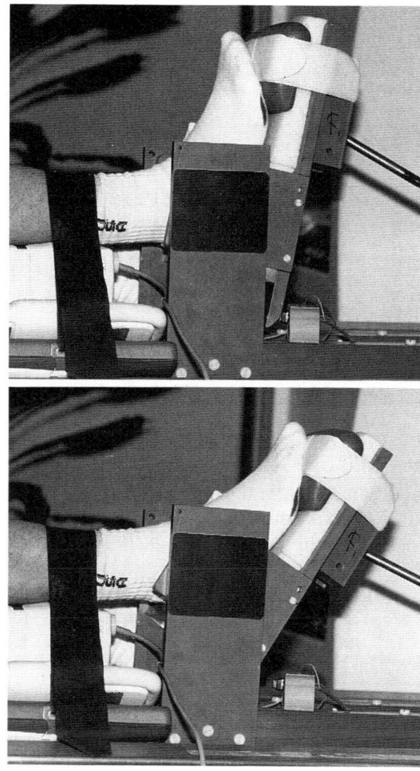

a

b

Fig. 10.5 a, b. Ankle flexion exercise. This movement is motored almost exclusively by the calf muscles. The surface coil registers the metabolite balance before, during, and after exercise, continuously and noninvasively

foot fixed in an special nonferromagnetic ergometer for use inside the magnet (Fig. 10.5). An exercise can thus be performed by flexing the ankle against a load that can be varied.

The exercise protocols used by different authors is quite variable (Taylor et al. 1986; Radda 1988; Argov et al. 1987; Achten et al. 1990). Spectra can be recorded during discrete levels of steady-state work in which each level of work must be performed for some minutes to assure the steady state. A far more practical approach, however, consists of a ramp exercise protocol in which the workload is increased by a fixed amount every minute (March et al. 1989). During each level of work, 1-min spectra are recorded every 30 s. dynamics of recovery after the last exercise spectra are recorded every 30 s. Figure 10.6 shows the total series of spectra recorded from the calf muscle of an endurance-trained athlete, before, during, and after exercise. Figure 10.7 compares the spectrum before exercise with the last spectrum during exercise. The pHi has decreased from 7.01 to 6.75. The relative PCr content dropped from 66% to 25% of the total visible ^{31}P, and the Pi content rose accordingly from 9% at rest to 50% at the end of exercise. The diagram in Fig. 10.8 demonstrates the changes of the relative concentrations of the metabolites over time; ATP remained remarkably constant in this well-trained individual.

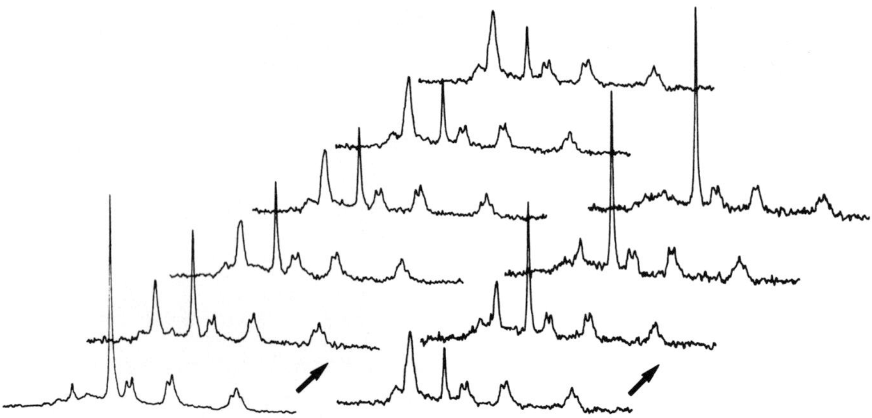

Fig. 10.6. Sequence of spectra recorded from the calf muscles of an endurance-trained subject. *Bottom left*, spectrum before exercise; this is followed to by five 1-min spectra during increasing workload. It can be seen clearly that the area of the PCr peaks decreases while that of the Pi peaks increases correspondingly. The area under the ATP peaks remains remarkably constant in this trained subject. *Bottom right*, last spectrum during exercise; this is followed by the first three spectra in the recovery phase. The PCr area increases rapidly while Pi disappears almost completely

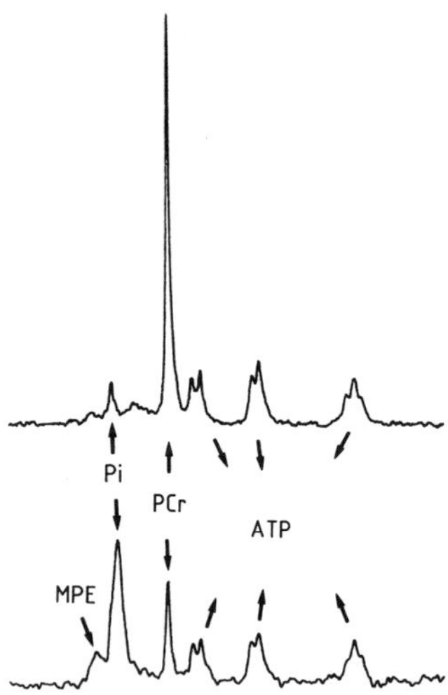

Fig. 10.7. Comparison between the rest spectrum and the spectrum acquired during the last minute of exercise (70% of MVC). The large increase in Pi and decrease in PCr are evident. Note also the prominent sugar phosphate peak in the end of exercise spectrum, indicating glycolysis

10.4 Use of ^{31}P MRS in Sports Physiology

10.4.1 The Energy Sources: A Reminder

The mechanism of muscular contraction was first described by Huxley and Gordon (1982). This is the result of the relative sliding of actine and myosine filaments. The driving force of contraction and relaxation comes from

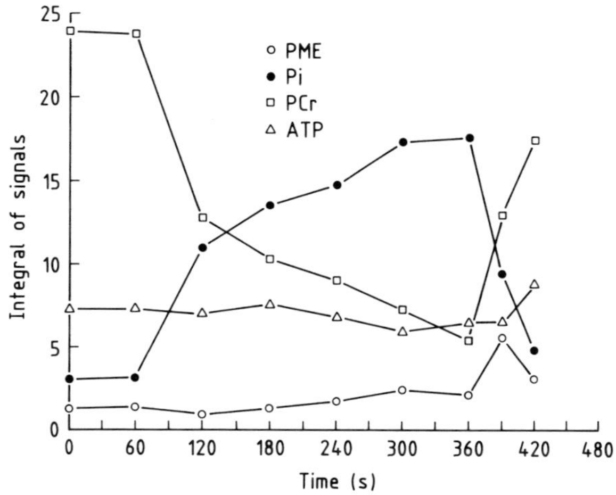

Fig. 10.8. Relative concentration changes and the pH changes calculated from the spectra of the calf muscles shown in Fig 10.7. The workload is expressed as percentage of MVC. Note that the changes of Pi and PCr are a mirror image of each other. ATP remained more or less constant. pHi decreased from 7.01 before to 6.75 after exercise

hydrolysis of ATP (ATP is used by cross-bridges, but also by the calcium pump):

$$ATP + H_2O \Leftrightarrow ADP + Pi$$

To ensure muscular contraction, it is imperative to keep a relatively constant level of ATP in the muscle fibers. This is possible by the breakdown of PCr:

$$PCr + ADP \Leftrightarrow Cr + ATP$$

The splitting of PCr is a very fast process and absolutely necessary to bridge the time that other mechanisms need to resynthesize ATP. This makes PCr an important instant store of high energy phosphate in muscle. The PCr concentration constitutes 18–20 mmol/kg wet weight, compared to the ATP concentration of 4–5 mmol/kg wet weight (Karlsson 1971).

An other way to resynthesize ATP is through its formation from two molcules of ADP, a reaction catalyzed by adenylate kinase (myokinase):

$$ADP + ADP \rightarrow ATP + AMP$$

Adenosine monophosphate (AMP) is removed by deamination with the formation of inosine monophosphate (IMP) and amonium ion (NH^+):

$$AMP + H_2O + H^+ \rightarrow IMP + NH_4^+$$

ATP can be very quickly replenished in contracting muscle by the creatine kinase reaction and in extreme conditions also by the myokinase reaction. The supply of substrate for these reactions is limited, however, and the most

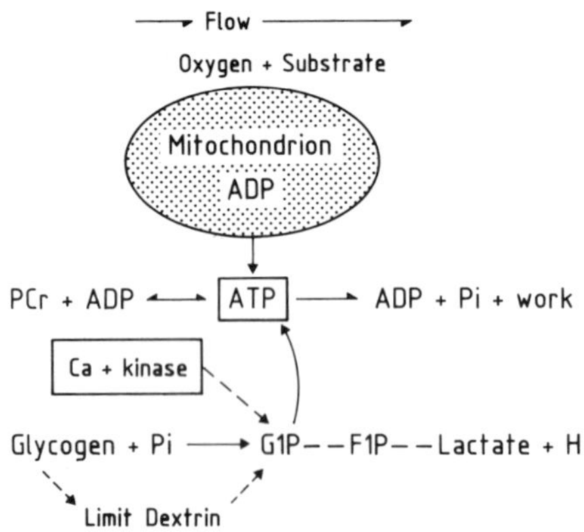

Fig. 10.9. Metabolic flowchart showing the major pathways to conserve ATP during exercise in the muscle

important resynthesis of ATP for exercise taking more than a few tens of seconds comes from anaerobic glycolysis and oxidative phosphorylation.

The most economic means of ATP resynthesis is the burning of fat, this being the largest energy reserve of the body. The amount of ATP synthesized out of free fatty acids (FFA) is huge. Nevertheless the use of FFA for ATP synthesis is limited because of limitations in reaction rate and the speed of substrate delivery to the muscle cells. Oxidation of carbohydrates also produces a large amount of ATP, and its production rate is much higher. The glycogen content of muscle is limited and thus not sufficient for extremely long efforts, this in spite of many dietary manipulations to increase body glycogen reserves. The average glycogen content of human muscle is approximately 90m*M* glycosyl units per kilogram tissue (Jue et al. 1989). If the needs of muscle for ATP exceeds the rate of combustion of FFA and glycogen with oxygen, ATP is resynthesized by anaerobic glycolysis. This pathway is much less efficient in the amount of ATP produced per unit of substrate (glycogen) than the aerobic pathways. During anaerobic exercise, lactate, $HPO_4^=$ and H^+ production by anaerobic glycolysis ultimately limits the effectiveness of the pathway, and exercise is stopped due to ATP exhaustion even before glycogen stores are depleted.

Figure 10.9 summarizes the ways in which ATP is used and supplied by oxidative, glycolytic, and other reactions (Radda 1988). These different ATP sources are enrolled one after the other in the event of muscular activity (Newsholme 1988).

10.4.2 Quantification with ^{31}P MRS

^{31}P MRS can be a valuable instrument for quantifying the activity of these energy sources. Chance et al. (1987) developed a quantitative approach to characterize the transfer function between energy expenditure, measured as work, and the metabolic rate, determined by the level of the key substrates for metabolic control. These authors used low-level steady-state exercise in which [ATP] was constant. It is useful to apply Michaelis-Menten enzyme kinetics (Lehninger 1982) to describe the transfer function as the rate of reaction of enzymes controlled by substrate levels in the form of a rectangular hyperbola:

$$V/V_m = 1/(1 + K_m/S)$$

where V is the observed velocity, V_m is the maximum velocity, K_m is the affinity constant, and S is the substrate which is regulating the enzymatic activity. Many substrates may be involved, and the equation can be rewritten as:

$$V/V_m = 1/(1 + k_1/\mathrm{ADP} + k_2/\mathrm{Pi} + k_3/\mathrm{O_2} + k_4/\mathrm{NADH})$$

During low-level steady state work where the assumptions, pH = 7.0, excess Pi, O_2, NADH are valid, the previous equation becomes:

$$V/V_m = 1/(1 + K_1/\mathrm{ADP})$$

ADP is the major regulating substrate in skeletal muscle. Using the rate equation for the creatine kinase reaction:

$$K_m = ([\mathrm{ATP}][\mathrm{Cr}])/([\mathrm{H}^+][\mathrm{ADP}][\mathrm{PCr}])$$

with [ATP] constant and [PCr] + [Cr] = [PCr] + [Pi], we can write this as:

$$[\mathrm{ADP}] = \mathrm{K} * [\mathrm{Pi}]/[\mathrm{PCr}]$$

and the Michaelis-Menten rate equation can be simplified as:

$$V/V_m = 1/[1 + K_x/(\mathrm{Pi/PCr})]$$

where K_x is a constant. Thus, for skeletal muscle the regulating variable is the free concentration of ADP, which is reflected in the MR spectra by Pi/PCr. For low V/V_m the approximation is:

$$V/V_m = 1.7\ \mathrm{Pi/PCr}$$

For steady-state exercise at low work levels, this equation can be used to quantify the aerobic function in skeletal muscle, as demonstrated in Fig. 10.10. Increasing work levels rapidly result in a drop in pH, however, and the onset of non-steady-state metabolism.

Another way to quantify aerobic function is described by Achten et al. (1990). The recovery rate of PCr is biphasic (Argov et al. 1987; Taylor et al. 1983); a first, rapid phase is due to the high anabolism of ATP after exercise, resulting from the high oxidative metabolism at that moment. Because the actual needs for ATP are low immediately after exercise, the ATP produced by aerobic means is used to rephosphorylate creatine to PCr. Spectroscopically, this is reflected by a rapid increase in PCr (Peter et al. 1972) and decrease in the Pi signal (Fig. 10.11). Therefore the recovery rate of Pi is a good measure for the mitochondrial or oxidative function immediately after exercise (Achten et al. 1990).

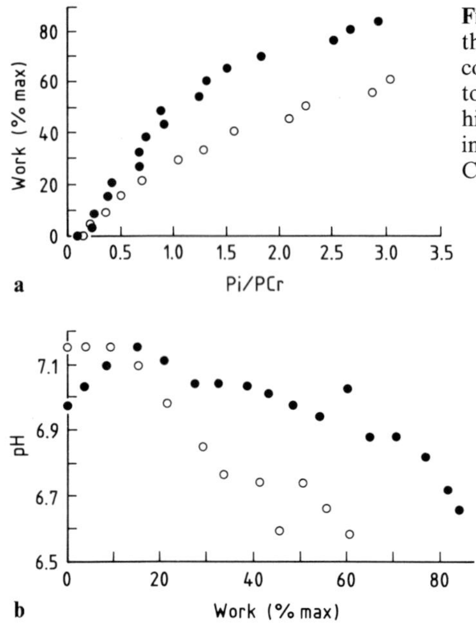

Fig. 10.10. Values for Pi/PCr ratio and for the work of the wrist flexor muscles of two control subjects during a ramp exercise protocol. Note the difference in the Pi/PCr at the higher work levels as well as the greater drop in pH in one of the subjects. (Courtesy of B. Chance)

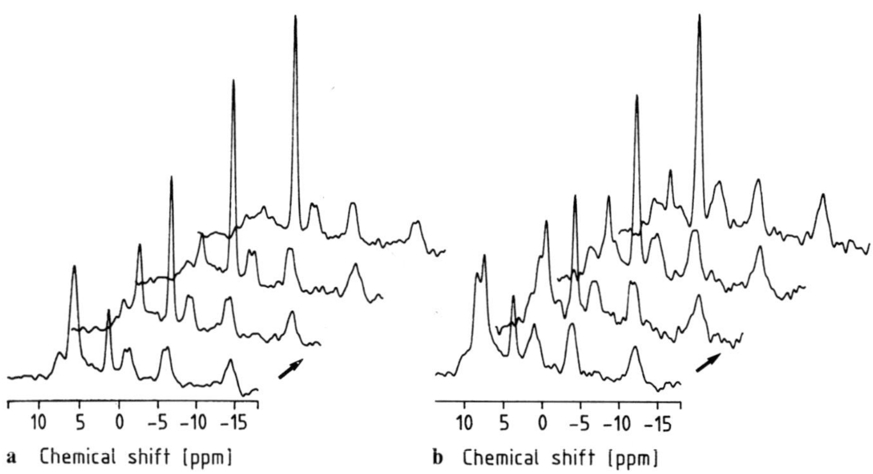

Fig. 10.11. The recovery spectra at 30, 60, and 90 s after exercise, superimposed on the spectra at end of exercise for an endurance-trained subject (*A*) and for a sedentary subject (*B*). The recovery of Pi is faster in the endurance-trained subject

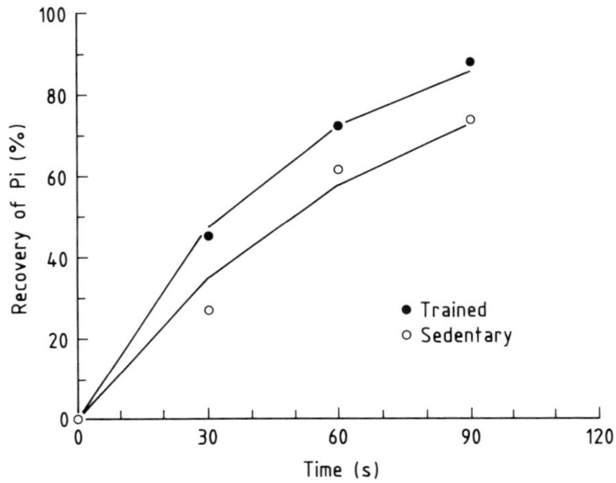

Fig. 10.12. Recovery curves for Pi for an endurance-trained and for a sedentary subject. The initial recovery rates calculated from these curves using a monoexponential model are 2.13% s^{-1} for the former and 1.47% s^{-1} for the latter, indicating a 1.48 times higher aerobic function at the end of exercise

These authors use a ramp exercise protocol to examine the performance of human calf muscle. In order to stress the calf muscles, the subjects were asked to flex the ankle (plantar flexion) with a frequency of one per second, against a load that increased every minute for 5 min. Spectra were acquired before exercise, every minute during exercise, and every 30 s during first 5 min of recovery. The recovery of the Pi peak in percent at time t after the end of exercise is calculated as follows:

$$R(t) = 100\,[\mathrm{Pi}\,(0) - \mathrm{Pi}\,(t)]\,/\,\mathrm{Pi}\,(0)$$

for $t = 0, 30, 60, 90$ s. Pi (0) is the relative peak integral of the Pi peak at the end of exercise, whereas Pi (t) is the relative peak integral of Pi at time t after the end of exercise. To these data, a monoexponential function of the form:

$$R(t) = 100\,[1 - \exp(-t/T)]$$

can be fitted with a nonlinear regression algorithm, where t is the time after the end of exercise, and T the exponential time constant associated with Pi recovery. The slope at $t = 0$, S_0 of these curves:

$$S_0 = (\mathrm{d}R(t)/\mathrm{d}t)_{t=0} = 100/T$$

is the initial rate of Pi recovery expressed as % s^{-1} and is a measure for the aerobic metabolism at the end of the exercise. The decrease in pH during exercise is a measure for the degree of anaerobic glycolysis, and the ratio Pi/PCr reflects the activity of the creatine kinase enzyme system. All these parameters allow the noninvasive quantification of exercise performance in vivo in a matter of minutes.

Figure 10.12 shows the difference that exists in the aerobic function of the calf muscles of a sedentary subject and an endurance-trained athlete at the end of an exhaustive exercise. The initial rate of Pi recovery is about 50% higher in the trained subject. The pHi at the end of exercise was 6.75 in the calf muscles of the endurance-trained athlete and 6.23 in the muscles of the sedentry subject, accounting for the greater dependence on anaerobic glycolysis for ATP production for the same relative ammount of work.

It is clear from this that training effects can be quantified with ^{31}P MRS, and this makes the technique an ideal tool to study fundamental physiological adaptation mechanisms of the energy system of muscle in response to exercise (Peter et al. 1972; Achten et al. 1988; Chance et al. 1983; McCully et al. 1988 a, b; Molé et al. 1985; Park et al. 1987 a).

10.4.3 The Mechanism of Fatigue

^{31}P MRS has already been used extensively to study several other metabolic and physiological events related to skeletal muscle activity. A current topic in this research is the phenomenon of fatigue. In 1977, Dawson et al. (1978) found that the decrease in force development in isolated anaerobic frog muscles undergoing repeated isometric contraction at 4 °C is approximately proportional to the rise in [ADP] and [H^+], and that this dependence of isometric force development on [ADP], [Pi], or [H^+] could arise from product inhibition of the contractile system and from interference with the activation of contraction.

More recently, Carson et al. (1987) demonstrated that the constant relationship between metabolites, H^+, and $H_2PO_4^-$ supports the view that high-energy phosphate metabolites, pH, and $H_2PO_4^-$ are causative factors in fatigue regardless of muscle type or form of exercise. However, these metabolic changes produced by high-intensity exercise contrast with those found by low-intensity exercise, where there is minimal change in PCr, H^+, and MVC but marked reduction in twitch tension. These authors conclude that low-intensity exercise appears to cause "low-frequency" fatigue by inhibition of exitation contraction coupling, while high-intensity exercise causes fatigue by changing metabolite levels.

10.4.4 MRS in Muscle Injury

McCully et al. (1988 c) showed that elevated Pi/PCr and decreased ATP / (total ^{31}P) can occur in normal subjects with muscle fiber injury. Evidence of mitochondrial alterations have been reported subsequent to muscle injury and could account for this ^{31}P MRS result. Muscle edema occurs as a result of injury, and the increased extracellular fluid might contain elevated levels of Pi. This would result in an increase in the Pi signal relative to PCr and ATP signals. However, further work needs to be done to clarify the responsible mechanisms.

10.4.5 Muscle Fibers and MRS

It is common knowledge that most human skeletal muscles are built up of two main types of muscle fibers: type I, or aerobic fibers, and type II, or glycolytic

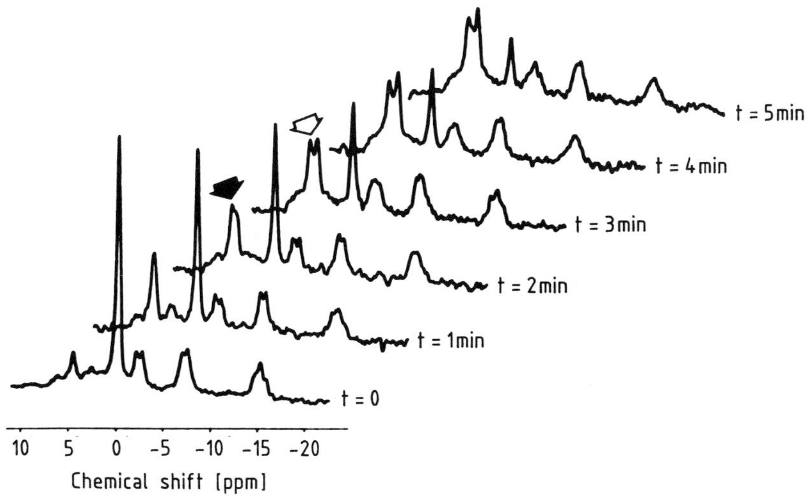

Fig. 10.13. Sequence of spectra from the calf muscle of a sedentary subject. *Bottom left*, spectrum during increasing workload. Note the broadening of the Pi resonance from the 2nd min of exercise (*closed arrow*) and a complete splitting from the 3rd min of exercise on (*open arrow*). This pH compartmentation is due to the differential recruitment of different muscle fibers

fibers. Substantial controversy exists whether the ratio of the number of fibers in muscle is more an inherited or more an acquired property (Peter et al. 1972).

Recently, Park et al. (1987b) observed a differential recruitment of oxidative and glycolytic fibers in wrist flexor muscle in four out of ten untrained men by monitoring the phosphate metabolites with MRS. Achten et al. (1990) showed that ^{31}P MRS is an excellent noninvasive procedure to quantify aerobic function in both fiber types simultaneously. In the calf muscles of seven sedentary subjects, exhaustive exercise resulted in clear splitting of the Pi resonance (Fig. 10.13), indicating pH compartmentation. From the chemical shift of the Pi peaks relative to PCr, the mean pHi at the end of exercise was 6.92 ± 0.05 in one compartment and 6.23 ± 0.15 in the other. The aerobic function at the end of exercise, expressed as the inital rate of recovery of the Pi peak, was $2.49 \pm 0.17\%\ s^{-1}$ in compartment I, almost three times higher than in compartment II, where the initial note of recovery was $0.87\%\ s^{-1}$. The authors explained this phenomenon by postulating that different metabolic behaviors in type I and type II fibers of human calf muscle in response to exhaustive exercise are observed.

References

Achten E, De Meirleir K, van Cauteren M, Osteaux M (1988) In vivo magnetic resonance spectroscopy of muscle metabolism. Belg Tijdschr Radiol 71:255–257

Achten E, van Cauteren M, Willem R, Luypaert R, Malaisse WJ, van Bosch G, Delanghe G, DeMeirleir K, Osteaux M (1990) ^{31}P NMR spectroscopy and the metabolic properties of different muscle fibers. J Appl Physiol (in press)

Argov Z, Maris J, Damico L, Koruda M, Roth Z, Leigh JS, Chance B (1987) Continuous graded steady state muscle work in rats studied with in vivo 31 phosphorous magnetic resonance. J Appl Physiol 63:1428–1433

Arnold DL, Taylor DJ, Radda GK (1985) Investigation of human mitochondrial myopathies by phosphorous magnetic resonance spectroscopy. Ann Neurol 18:189–196

Carson PJ, Moussavi RS, Miller RG, Weiner WM (1987) Evidence for a constant relationship between fatigue, high-energy phosphates and pH in different muscle types and in varying forms of exercise. SMRM, 6th annual meeting and exhibition, New York, p 583 (Book of abstracts, vol 2)

Chance B, Sapega A, Sokolow D, Eleff S, Leigh JS, Graham T, Armstrong J, Warnell R (1983) Fatigue in retrospect and prospect: 31P NMR studies of exercise performance. In: Knuttgen HG, Vogel JA, Poortmans J (eds) Biochemistry of exercise. Human Kinetics, Chicago, pp 895–908 (International series in sport sciences, vol 13)

Chance B, Nioka S, Leigh JS Jr (1987) Metabolic control principles: importance of the steady state reaffirmed and quantified by ^{31}P MRS. In: Oxygen transport and utilization. Society of Critical Care Medicine, Fullerton, pp 215–228

Dawson MJ (1982) Quantitative analysis of metabolite levels in normal human subjects by 31-P topical magnetic resonance. Biosci Rep 2:727–733

Dawson MJ, Gadian DG, Wilkie DR (1978) Muscular fatigue investigated by phosphorus magnetic resonance. Nature 274:861–866

Gadian D (1982) Nuclear magnetic resonance and its applications to living systems. Oxford University Press, pp 1–41

Grisogono V, Yaffé M (1986) In: Helal B, King JB, Grange WJ (eds) Sports injuries and their treatment. Chapman and Hall, London pp 1–32

Huxley AF, Gordon AM (1962) Striated patterns in active and passive shortening of muscle. Nature 193:280

Jue T, Rothman L, Tavitian A, Shulman RG (1989) Natural-abundance 13C NMR study of glycogen repletion in human liver and muscle. Proc Natl Acad Sci USA 86:1439–1442

Karlsson J (1971) Lactate and phosphagen concentrations in working muscle of man. Acta Physiol Scand Suppl 358:1–72

Lehninger AL (1982) Enzymes. In: Anderson S, Fox J (eds) Principles of biochemistry. Worth, New York, pp 207–247

March GD, Thompson RT, Paterson DH, Driedger AA (1989) Assessment of oxydative metabolism in normal and phosphorilase deficient patients using a ramp exercise protocol and P31 NMR spectroscopy. SMRM, 8th annual meeting and exhibition, 12–18 Aug 1989. Works in progress. p 1089

McCully KK, Kent JA, Chance B (1988a) Application of 31P Magnetic Resonance Spectroscopy to the study of athletic performance. Sports Med 5:312–321

McCully KK, Argov Z, Boden BP, Brown RL, Bank WJ, Chance B (1988b) Detection of muscle injury in humans with 31P magnetic resonance spectroscopy. Muscle Nerve 2:212–216

McCully KK, Argov Z, Boden, BP, Brown RL, Bank WJ, Chance B (1988c) Detection of muscle injury in humans with 31-P magnetic resonance spectroscopy. Muscle Nerve 2:212–216

Molé PA, Coulson RL, Caton JR, Nichols BG, Barstow TJ (1985) In vivo 31P-NMR in human muscle: transient patterns with exercise. J Appl Physiol 59:101–104

Newman RJ, Bore PJ, Chan L (1982) Nuclear magnetic resonance studies of forearm muscles in Duchenne dystrophy. Br Med J 284:1072–1074

Newsholme EA (1988) Basic aspects of metabolic regulation and their application to provision of energy in exercise. In: Poortmans JR (ed) Principles of exercise biochemistry. Karger, Basel, pp 40–77 (Medicine and sport science, vol 127)

Park JH, Brown RL, Park CR, McCully KK, Cohn M, Chance B (1987a) High energy phosphate metabolism in athletes and normal subjects during prolonged exercise. SMRM, 6th annual meeting and exhibition, New York, p 1041 (Book of abstracts, vol 2)

Park JH, Brown RL, Park CR, McCully K, Cohn M, Haselgrove J, Chance B (1987b) Functional pools of oxidative and glycolytic fibers in human muscle observed by 31P magnetic resonance spectroscopy during exercise. Proc Natl Acad Sci USA 84:8976–8980

Peter JB, Barnard RJ, Edgerton VR, Gillespie CA, Stemple KE (1972) Metabolic profiles of three fiber types of skeletal muscle in guinea pigs and rabbits. Biochemistry 11::2627–2633

Petroff OAC, Prichard JW, Behar KL, Alger JR, den Hollander JA, Shulman RG (1985) Cerebral intracellular pH by 31 P nuclear magnetic resonance spectroscopy. Neurology 35:781–788

Radda G (1988) Clinical applications of magnetic resonance spectroscopy. In: Budinger TF, Margulis AR (eds) Medical magnetic resonance: a primer. SMRM, pp 295–307

Taylor DJ, Bore P, Styles P, Gadian DG, Radda GK (1983) Bioenergetics of intact human muscle. A 31 phosphorous nuclear magnetic resonance study. Mol Biol Med 1:77–94

Taylor DJ, Styles P, Matthews PM, Arnold DA, Gadian DG, Bore P, Radda GK (1986) Energetics of human muscle: exercise induced ATP depletion. Magn Reson Med 3:44–54 (1986)